INTERPRETATION OF
GEOLOGICAL STRUCTURES
THROUGH MAPS

AN INTRODUCTORY PRACTICAL MANUAL

INTERPRETATION OF GEOLOGICAL STRUCTURES THROUGH MAPS

DEREK POWELL

Longman Scientific & Technical,
Longman Group UK Limited,
Longman House, Burnt Mill, Harlow,
Essex CM20 2JE, England
and Associated Companies throughout the world.

Copublished in the United States with
John Wiley & Sons, Inc., 605 Third Avenue, New York, NY 10158

British Library Cataloguing in Publication Data
Powell, Derek
    Interpretation of geological structures through maps: An introductory practical manual.
I. Title
912

ISBN 0-582-08783-X

Library of Congress Cataloging-in-Publication Data
A catalogue record for this book is available from the Library of Congress.

ISBN 0470-21822-3 (USA only)

Set in Garamond 10 on 12pt

Printed in Hong Kong
SK/01

CONTENTS

PREFACE

In the same way that topographic, road, and rail maps provide us with information concerning the nature of the land surface and the location of man-made features, geological maps contain data which allow an understanding of the distribution of rocks that make up the crust of the Earth and the orientation of structures they contain. Unlike ordinary maps, however, geological maps include information which allows us to assess not only the location of particular rocks and the areas they cover, but also their underground extent and their geological history.

Geologists construct geological maps by making observations of the nature of rocks exposed at the surface of the Earth, in drill-holes and mine shafts, and recording these on topographic maps and/or aerial photographs. In doing this they plot the locations of contacts between different rock types and measure the attitudes of these and other planar and linear features within rocks. From such information geologists can predict the shapes of rock formations at depth—formations which, in some instances, may contain gold, oil or gas, etc.

Although geological maps are two-dimensional, knowledge of how to interpret them permits an understanding of the extent in three dimensions of the geological features they show, i.e. both below ground and, before they were eroded away, above ground level. The ability to employ geological maps successfully in this way depends not only on interpretation of direct measurements of the attitudes of planar and linear geological features, but also on an understanding of the relationships between the shapes of bodies of rocks, as seen on maps, and the shape of the ground surface (i.e. the **topography**).

Map interpretation is vital to all who wish to understand geological processes fully, but it confronts many students and practitioners of geology with difficulties. This is because it is necessary to gain a three-dimensional picture in the mind and eventually on paper, from data that are presented in two dimensions, i.e. as a geological map—a task that is not often met in other subjects. For a few, this ability is gained quickly but, for most of us, it takes longer and is very much a matter of practice making perfect. Consequently this manual attempts to give an appreciation of the basic problems and techniques involved in unravelling the geological structure of an area from data presented as a map. Further, by pursuing some problems of analysis through use of structure contours, it attempts to encourage the reader to develop the ability to manipulate three-dimensional data.

Some teachers of geology are opposed to using structure contours as an introduction to the interpretation of geological maps on the grounds that (a) they can become unrealistic and (b) they are seldom used in the interpretation of 'real' geological maps. The first criticism is true of some books on map interpretation, but nevertheless their use still remains the only clear, and least qualitative, method of introducing the three-dimensional problems involved in map interpretation. The second criticism is not true, as structure contours and their derivatives are used widely to solve problems in the oil and mining industries as well as in academic research.

This manual provides elementary instruction and exercises which are intended (a) to encourage development of the ability to undertake three-dimensional analysis and (b) to illustrate some of the basic techniques and problems involved in the three-dimensional analysis of geological maps. It places particular emphasis on the geometries of bodies of rock and structures, and their relationships in time. It is not intended as a reference book for all the techniques involved in map analysis, nor for all types of structures, but rather it offers a collection of explanations, examples and exercises which are aimed at allowing a student of geology to develop the ability to think in three dimensions and quickly derive a three-dimensional appreciation of data recorded on geological maps.

For each exercise in the manual, a solution (together with discussion) is given in Chapter 15. This will allow you to check your analyses and assess whether or not you are gaining a full appreciation of the problems posed, the techniques involved, and possible interpretations.

Amongst recent books dealing with the interpretation of geological maps in more general terms than this manual, those by Butler and Bell (1988) and Maltman (1990) are recommended. They do not however emphasise to such an extent the geometry of structures and their analysis. For the reader interested in acquiring more knowledge of geological structures, an excellent introduction is provided by Davis (1984).

Butler B C M, Bell J D 1988 *Interpretation of Geological Maps.* Longman Scientific and Technical Harlow

Davis G H 1984 *Structural Geology of Rocks and Region* John Wiley & Sons New York

Maltman A 1990 *Geological Maps: An Introduction* Open University Press Milton Keynes

Derek Powell
Royal Holloway & Bedford New College
(University of London)
May 1991

ACKNOWLEDGEMENTS

My thanks are due to the many past undergraduates of Royal Holloway and Bedford New College (University of London) and one of its predecessors, Bedford College, who have, however unknowingly, taught me a lot about geological structures and map interpretation. Postgraduate demonstrators in my field and practical classes also deserve my appreciation. They have offered encouragement and advice without which this manual would not have been finished.

My wife, Pamela, has exercised considerable patience through the latter stages of completing the manuscript and has given much helpful advice and criticism in its editing.

INTRODUCTION TO GEOLOGICAL STRUCTURES

Over the past four thousand million years of Earth history the crust of the Earth has, to varying degrees, been mobile. As a consequence many of the rocks that we now see at, or near, the surface, no matter what their origin, have been squashed, stretched or fractured; they have been deformed. Thus all geologists need to be aware of the nature and effects of the various **structures** in rocks which have allowed displacements of the Earth's crust and changes in the shapes of bodies of rock. Although structural geology is the branch of the Earth sciences directly concerned with the description, analysis and origin of those phenomena in rocks which are the result of such **deformation**, a basic understanding of rock structures is essential to all students of geology.

Deformation arises because large parts of the Earth underlying the oceans and continents (known as lithospheric plates) have been moving relative to each other throughout much if not all of geological time. These differential plate movements involve distances of hundreds to thousands of kilometres at rates of up to 12 cm per year so that in some instances, for example, continental areas have moved from polar to equatorial regions and beyond. Today, as in the past, such movements are responsible for earthquake activity where plates are either colliding, sliding past each other, or pulling apart.

Whilst relatively slow, plate movements persist for millions of years and generate **stresses** that lead to both overall **compression**, where plates collide, and overall **extension**, where they stretch or break apart. The rocks comprising the crust respond to such stresses by undergoing changes of shape (**strain**) so that on scales ranging from hundreds of kilometres to the sub-microscopic, various types of **geological structures** are developed which provide a record of the type of deformation suffered. Structures may be pervasive when rocks are capable of flow when deformed, or may involve discrete zones of deformation. When rocks are brittle they fracture as soon as levels of stress exceed the strength of the rocks involved. Figure 1 illustrates three of the main types of structure produced by compression. It shows that, in response to directed compressive stress, rocks may fracture to produce faults, they may fold, or by changes in the shapes of constituent grains and crystals, by solution of material together with the growth of new crystals, they may undergo an overall homogeneous shape change with the production of planar fabrics; these are known as cleavage or schistosity.

As a result of plate movements, most rocks making up the Earth's crust have been or will be deformed to some degree. Consequently an understanding of the characteristics, origin and significance of structures is important to our understanding of geological processes as a whole and the history of the Earth

FIG. I Basic types of geological structure

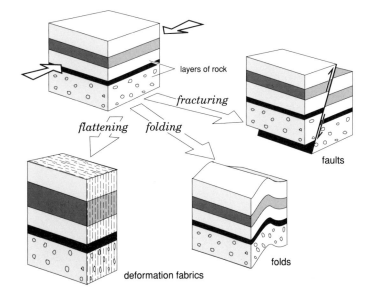

over the past 4000 million years. For example, all of the major mountain belts on Earth are the result of a complex interaction of plate extension and plate collision and as a result they contain, amongst others, structures such as folds and faults which record the type of strain suffered and the history of deformation.

We are therefore concerned with the nature and orientation of stress systems that operated in the past, together with the strain features such as faults, folds and planar and linear rock fabrics, that record the response of rocks to these stresses. Consequently we require an appreciation of the geometry and significance of rock structures, and the small- and large-scale processes that give rise to them.

A basic tool of the geologist is the **geological map**. On these are recorded the distribution of different rock types and their contacts; the attitudes of internal layering and boundaries; and the deformation structures the rocks may contain. The geological map provides information from which both the sub-surface extension of geological features seen at ground level can be extrapolated and their original, above-ground (pre-erosion) extent can be assessed. By geometric, three-dimensional analysis of the shapes and attitudes of geological features and their inter-relationships, we can achieve an understanding of not only the types of geological phenomena present, but also, in the case of structures, the generative stress systems. Further, we can place the development of different rock types, geological features and structures in a time framework and thus deduce the geological history of an area and its tectonic setting, i.e. whether it relates to an extensional, compressional or stable environment, or at different times two or three of these. As well as being important for the appraisal of economic potential, e.g. in locating ore-bodies, oil reservoirs, etc., and assessing their shape and size, such knowledge also helps in our appreciation of geological processes as a whole and helps us ultimately, in conjunction with other geological information, to determine plate tectonic configurations that existed in the past.

TYPES OF PLANAR GEOLOGICAL SURFACES

Analysis of geological maps requires an understanding of the origin and nature of the various rock types that make up the crust of the Earth and the planar and linear features developed in them. Planar and linear features arise principally through the following geological processes:

(i)   Accumulation over periods of time of sediments (with different compositions), in sub-aqueous (marine, lake or river) and sub-aerial (mainly desert or glacial) environments—**bedding planes, unconformities** (Figs 2, 3).

(ii) The intrusion of molten rocks into the crust and their solidification—**contacts** of igneous intrusions e.g. **dykes, sills, sheets,** etc. (Figs 2, 6).

(iii) The extrusion of molten rocks at the Earth's surface as volcanic lavas and ashes—**contacts** between **lava flows; internal flow fabrics; bedding planes in ashes,** etc. (Fig. 6).

(iv) The deformation of rocks—**folds of bedding,** etc.; **deformation fabrics,** e.g. **cleavage; fault** and **joint planes** (Figs 4 and 5).

2.1 PLANAR SURFACES ARISING FROM EROSION AND SEDIMENTATION

Figure 2, diagram **A**, illustrates the geology of a continental margin wherein continental crustal material, comprising an older basement complex intruded by igneous rock, together with a tilted, older sequence of sediments, is being covered by new sediments. Note that the contact between the new sediments and the older rocks is discordant (or disconformable) to the contacts between the different elements of the older complex; that is, it truncates them. Such discordant planar contacts, when produced by erosion and sedimentation, are known as **unconformities**. With the passage of time, a rise in sea level and erosion of the land causes the sea and the new sediments to transgress across the continental crust (diagram **B**); different types of sediments in the diagram are indicated by their different ornamentation and their contacts constitute further

FIG. 2 The development of bedding planes and unconformities

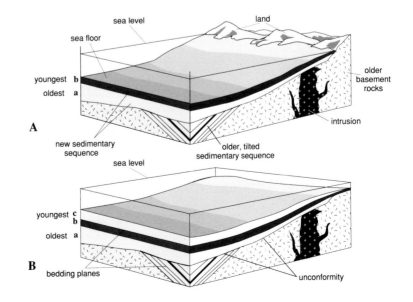

FIG. 3 Unconformities

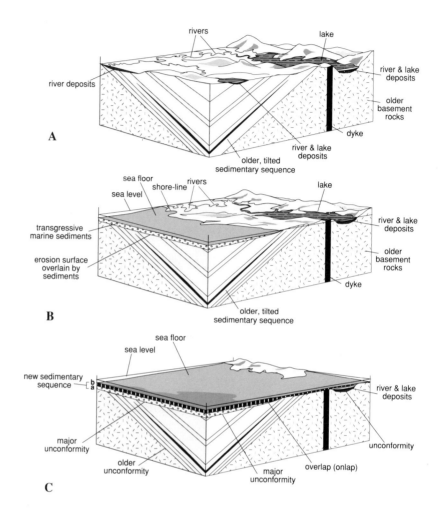

major planar surfaces known as **bedding planes**. Note that progressively younger sediments are deposited on top of older—c on b on a.

Unconformities provide important geological evidence for the passage of time and a key to the recognition of cycles of events in the geological evolution of an area. Thus in Fig. 2 the geological processes involved in the development of the basement complex—its intrusion by igneous material; deposition of the overlying, older sedimentary sequence and its subsequent tilting from its original sub-horizontal depositional attitude—all preceded formation of the unconformably overlying new, sedimentary sequence. They are thus older than deposition of the new, sedimentary sequence. By comparison the folding and faulting seen in Fig. 4 (diagram **B**) occurred after deposition of the new, sedimentary sequence and therefore constitute a separate, much younger set of events.

Unconformities can be widespread, developing over areas of thousands of square kilometres and therefore marking major changes in Earth history. Or they can be local, giving evidence of more restricted changes in geological processes. For example, in Fig. 3 (diagram A) uplift of the Earth's crust has caused erosion and formation of hills and valleys wherein river and lake sediments are deposited unconformably on an older basement complex, in local sedimentary basins. Note that the unconformities are local and at different heights.

In diagram **B**, further erosion has reduced the topographic relief and a rise in sea level has caused the sea to transgress the land. Marine sediments are deposited unconformably on the erosion surface. Continuing transgression (diagram **C**) carries marine sediments further across the older basement rocks and, in places, the river and lake sediments. A major unconformity is produced covering a large area and lying at approximately the same height. Note that, as transgression proceeds, younger deposits of the new sedimentary sequence (**b**) extend further than the older (**a**), a situation known as **overlap** (**onlap** in American literature).

Where we can work out the relative timing of geological events we can erect what is known as a **stratigraphic succession** for formation of the different rocks in an area. Thus for Fig. 3 the rocks can be arranged in order of their relative ages:

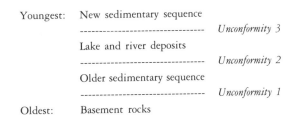

The relative age of the igneous intrusion is not, however, accurately known; it must be younger than the basement rocks that it intrudes and older than unconformity 2, but we do not, in the diagrams, see its relationships to the older sediments—it could be older or younger than these.

In the same way that we determine the stratigraphic succession given above, we can deduce the following time/event column for the area:

Youngest:	Deposition of new sedimentary sequence	(x)
	-- *Unconformity 3*	(ix)
	Erosion	(viii)
	Deposition of lake and river sediments	(vii)
	-- *Unconformity 2*	(vi)
	Tilting, uplift and erosion of older sedimentary sequence and basement	(v)
	Deposition of older sedimentary sequence	(iv)
	-- *Unconformity 1*	(iii)
	Erosion	(ii)
Oldest:	Formation of basement rocks	(i)

2.2 PLANAR SURFACES ARISING FROM CRUSTAL COMPRESSION

In Fig. 4 (diagram *A*), the same sedimentary basin and basement complex, seen in Fig. 2, undergo compression so that the rocks fold and are transported along a low-angled, compressional fault and the whole region is uplifted (diagram **B**). Development of the folds gives rise to curvi-planar geological surfaces, e.g. the folded unconformity and bedding planes, whilst the fault trajectory is itself curvi-planar. Where the rocks are folded upwards they are termed **antiforms** where downwards, **synforms**.

In diagram **C** uplift gives rise to erosion so that the deformed rocks become progressively dissected. Note that the crests and troughs of the folds in diagram **B** have a linear trend and that within the block as a whole there will be **lines of intersection** between the various planar elements. For example, such lines will occur where bedding planes in the older sediments run into the

FIG. 4 Structures produced by crustal compression

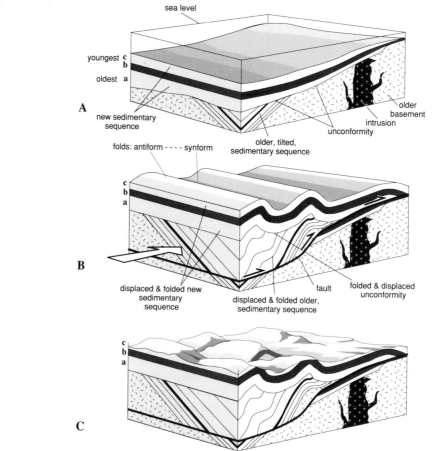

unconformity, and where the faults cut bedding planes and unconformities, etc. It will be shown later how such planar and linear features are important in the analysis of geological maps.

2.3 PLANAR SURFACES ARISING FROM CRUSTAL EXTENSION

When affected by extensional stresses the Earth's crust can fracture along planar **faults**, movement on which accommodates stretching. Thus in Fig. 5 (diagram **A**) slip on the fault causes extension: note also subsidence. Further extension can be accommodated by continued slip on the fault, but commonly further faults are formed as in diagram **B**. Clearly such faults are planar surfaces.

Although during extension the crust most commonly fractures along faults, in some instances the intrusion of thin, near-vertical sheets (**dykes**) of molten rock, as parallel regional swarms, accompanies such extension. Where dykes reach the surface of the earth volcanic fissure eruptions emit lavas and volcanic ashes. Figure 6 illustrates these processes and shows that, in such situations, as well as the planar and linear features already described, there will be planar contacts within and bounding the intrusive and extrusive igneous and volcanic materials and linear intersections between the dyke walls, unconformities and bedding planes, etc.

The foregoing discussion illustrates how various types of geologically defined planar surfaces originate. Clearly the expression of these, and other geological features, at ground surface and therefore on maps, depends on many factors. Amongst these are the shapes of geological surfaces, depth of erosion,

FIG. 5 Crustal extension by faulting

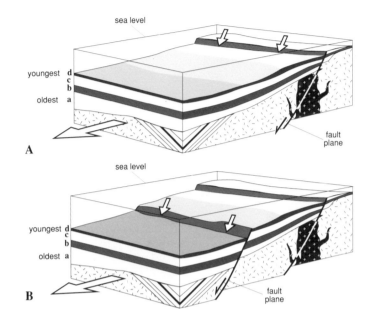

topographic relief, and amount of cover by soil and other superficial deposits. Consequently construction of most geological maps relies on incomplete first-hand data, i.e. limited direct observations of rock exposures at surface. Nevertheless, an understanding of the three-dimensional geometry of geological features allows controlled extrapolation between exposures and it is this understanding that is vital to both making geological maps and subsequently interpreting them. This manual is designed to encourage development of this understanding. Whilst it does not deal with techniques for map making nor with all the techniques for map interpretation, it offers an introduction to the three-dimensional problems faced and ways of analysing them. Emphasis is placed on the geometry of various types of structures but much of the reasoning underlying its interpretation is applicable to other geological phenomena.

FIG. 6 Intrusive and volcanic rocks resulting from crustal extension

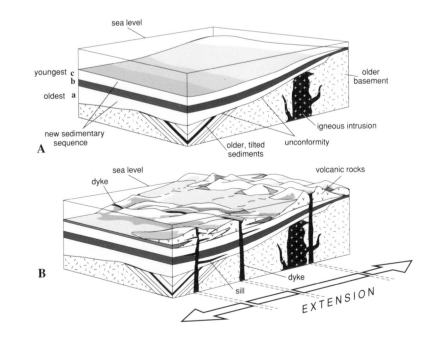

ANALYSIS OF PLANAR SURFACES

3.1 STRUCTURE CONTOURS

On maps the shape of the ground surface is reflected by topographic contours. These are the lines of intersection of imaginary horizontal planes of different heights above sea level (regarded as zero) with the ground surface. For convenience such planes are spaced at set intervals, i.e. 10, 25, 50 or 100 metres (or feet). As geological structures are often flat planar, or curvi-planar, surfaces, their shapes and attitudes can likewise be assessed by construction of contours on them. Figures 7 to 10 illustrate some of the basic principles and techniques involved in the derivation of such **structure contours**.

In Fig. 7 a layer of rock extends underground at an angle to the surface of

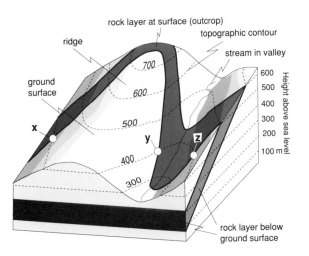

FIG. 7 Block diagram of an inclined layer of rock

the Earth. It intersects the topographic surface as shown by the curving shaded band (known as the **outcrop** of the rock layer). The trend and shape of this outcrop are dependent on both the inclination of the layer and the shape of the topographic surface. Thus the outcrop is V-shaped in the valley and Λ-shaped over the ridge. The height of the land surface above sea level (taken as zero), and its shape, are indicated by the topographic contours and because the latter intersect the rock outcrop we can determine the heights of both the top and bottom surfaces of the rock layer. Thus in Fig. 7 points **x**, **y**, and **z** on the top surface of the layer all lie at 400 m above sea level. Because they represent three separate points of the same height on the same geological surface, a line joining them together will define a 400 m contour on that surface. Such a line can be more readily visualised in Fig. 8 where the rocks overlying the layer have

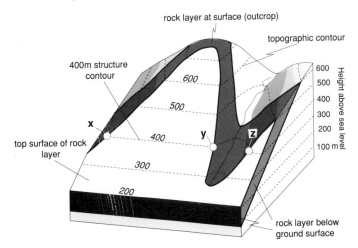

FIG. 8 Structure contours

9

FIG. 9 Extension of geological surfaces above ground level

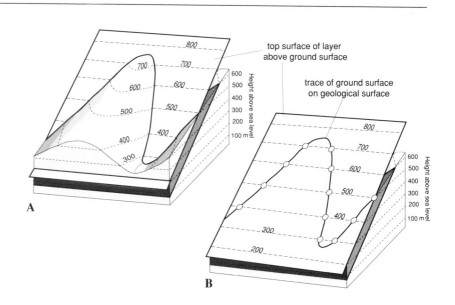

been removed and all lines of equal height (contours) on the top surface have been constructed.

Lines defined in this way are known as **structure contours** and are equivalent to topographic contours but drawn on a geological, rather than a land, surface. The structure contours in Fig. 8 are straight lines only because the geological surface is in this instance a flat, though inclined, plane.

In Fig. 9, diagram **A**, the top surface of the rock layer illustrated in Fig. 7 is drawn in to show how it would have extended before erosion cut the present land surface. Diagram **B** shows its full, pre-erosion, above and below ground, extent. We can deduce that this was the case by analysing a geological map of the block diagram (Fig. 10).

In Fig. 10 (diagram **A**) all the information in the block diagram has been projected vertically upwards onto an imaginary horizontal plane to produce a map (essentially a bird's eye view). Points **x**, **y** and **z**, as in Figs 7 and 8, define positions (circles) where the top surface of the rock layer lies at 400 m above sea level. The line drawn on the map connecting these positions is the 400 m

FIG. 10 Construction of structure contours

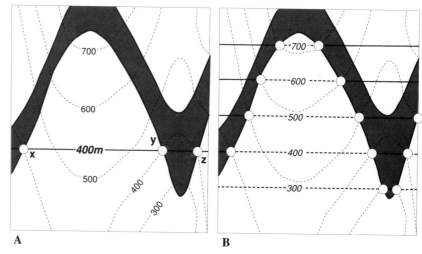

structure contour. In Fig. 10, diagram **B**, structure contours for heights of 700, 600, 500 and 300 m have been similarly constructed and where these lie underground they are shown as dashed lines, where above ground surface by solid lines. In the same way that topographic contours indicate the shape and slope of the land surface, the structure contours indicate the above- and below-ground position and attitude of the geological surface. The parallelism and even spacing of the structure contours indicate that the upper surface of the rock layer is a flat, but inclined, plane and the drop in height towards the bottom of the map shows that the surface is inclined (**dips**) in this direction. Notice that the outcrop of the rock layer in the valley is V-shaped in the direction of dip, over the ridge it is Λ-shaped. Understanding the significance of these shapes in relation to topography is extremely important in the analysis of geological maps, as will be shown later.

3.2 CROSS-SECTIONS

Although construction of structure contours on geological maps allows us to appreciate the shapes and attitudes of geological surfaces, it does not give us an immediate visual image of the structure. To achieve this we use a convenient and instructive method of illustrating the above- and below-ground extent of

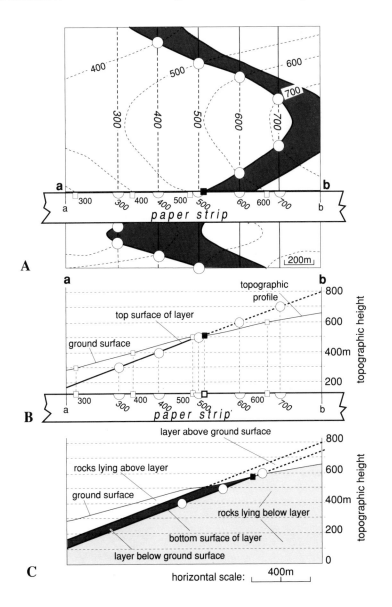

FIG. 11 Constructing cross-sections

geological surfaces, their attitudes and shapes—that is, construction of vertical cross-sections. Figure 11 illustrates the basic techniques used to do this.

In diagram **A**, for convenience of construction, the map (Fig. 10, diagram **B**) has been turned on its side. A line of section **a–b** is drawn at right angles to the trend of the structure contours and a paper strip placed along it. The points at which the section line crosses the topographic contours are noted on the strip of paper (white squares) and are transferred accurately to the vertical section as in **B**, so that a topographic profile can be drawn. Note that the vertical section (diagrams **B** and **C**) is constructed so that the vertical and horizontal scales are equal and the same as those of the map. As with topographic heights, the points at which the line of section cuts the structure contours are transferred (circles), as is the point at which the section line cuts the outcrop of the top surface of the rock layer (black square in **A** and **B**). Because all of these points must lie on the top surface of the rock layer, its position above and below ground level, and its inclination (its **dip**), can be constructed on the section by joining up all the transferred data points (diagram **B**). In diagram **C** the position of the bottom surface of the rock layer has been established using the same methods and we have therefore accurately recorded, on the section, the extent, position and attitude of the rock layer.

3.3 DIRECT MEASUREMENT OF DIP AND STRIKE

In the preceding section we have seen how the attitude of a geological surface can be determined from the relationships between outcrop and topographic contours. In practice, whenever possible, geologists measure the attitudes of geological structures at exposures of rock in the field while they are making geological maps. Such direct measurements can be made at rock exposures using a compass and clinometer. The former instrument allows direction to be measured; the latter, angle from horizontal.

The attitude of any plane can be defined in terms of the trend (azimuth) of a horizontal line drawn on it—its **strike**, and the azimuth and angle from the horizontal, of its direction of *maximum* inclination—its **dip** (Fig. 12). Note therefore that the strike of a surface is equivalent to the trend of a structure contour on the same surface (cf. Figs 12 and 8). The exposure illustrated in Fig. 12 is in a river bank at locality **A** in Fig. 13; it is viewed from the southwest. The contacts between the layers (**beds**) of sandstone and mudstone are inclined towards the left of the diagram (Fig. 12), i.e. towards the west, given that north in Fig. 13 is towards the top of the page.

The attitude of the bed surface shown can be measured in terms of the direction of its strike and its angle of dip. Note that the direction of dip, because it is the direction of *maximum* inclination, must, by definition, be at right angles to the direction of strike. In this case the surface strikes towards

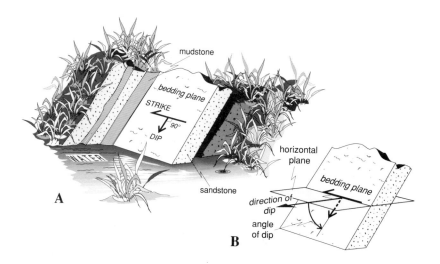

FIG. 12 Dip and strike of sandstone with mudstone interlayers, exposed in a river bank

FIG. 13 Geological maps showing location of exposure in Fig. 12

KEY TO SYMBOLS:

siltstones   ···200··· topographic contours

mudstones   geological boundaries

sandstones   stream

200m

due north (360) and dips to due west at 45°. Its attitude can be represented on the map as shown in Fig. 13, diagram **B**, whereon a dip and strike symbol is drawn accurately at the locality where the measurement was taken. Note that because the strike of any surface is by definition a horizontal line, in Fig. 12 it is parallel to the trace of the horizontal water surface on the bedding plane. Likewise water poured vertically down onto the bedding plane would run away down the dip of the surface.

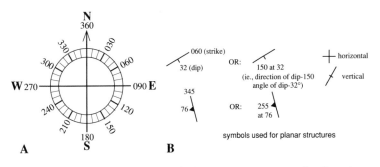

FIG. 14 Recording the attitudes of planar structures

symbols used for planar structures

When recording the attitudes of planar structures we refer directions to points of the compass as in Fig. 14, diagram **A**, and we make use of dip and strike symbols as in Fig. 14, diagram **B**. Note in **B** that we can refer to direction of strike and angle of dip OR to direction *and* angle of dip. Thus a strike of 060 and dip of 32° to the SE is the same as a dip *towards* 150 at 32°.

Analysis of the map (Fig. 13) using structure contours shows that the boundaries of the major rock units of the area dip consistently towards the west at 45° (Fig. 15). In the study of such geological maps, we can make use of both structure contours and direct measurements of dip and strike in order to assess fully the shapes and attitudes of geological surfaces.

3.4 EXAGGERATION OF SCALE

So far the cross-sections drawn to illustrate the attitudes of geological surfaces (Figs 11 and 15) have been constructed so that the vertical scales are equal to the horizontal. Such are **true scale** sections and therefore give accurate representations of the shapes of geological structures and attitudes of planar and linear features. In contrast, sections where the vertical scale is different from the horizontal cause distortion. Thus in Fig. 16, diagram **B**, where the vertical

FIG. 15 Relationships between dip and strike measurement and structure contours

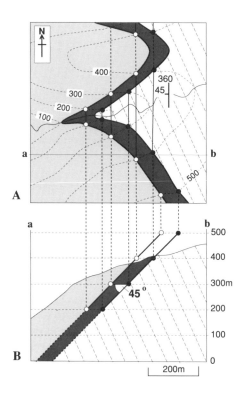

scale is twice the horizontal, angles of dip and layer thicknesses are increased, area is exaggerated, and shapes are changed. As well as being inaccurate representations of the geology, such sections can mislead our interpretations of the types of geological structures present; for example, note the differences in

FIG. 16 Exaggeration of scale in cross-sections

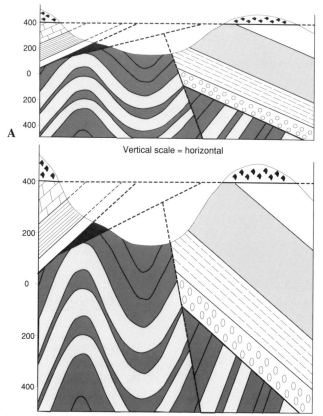

the shapes of the folds in the two sections. Consequently, whenever possible, cross-sections should be drawn to true scale.

3.5 CALCULATION OF TRUE AND APPARENT DIPS

Figure 12 shows that the true angle of dip of a geological surface (i.e. the maximum angle) is measured at right angles to the direction of strike. Therefore it follows, in relation to structure contours, that the direction of maximum dip will be at right angles to the trend of these contours. Because the spacing of structure contours is directly related to the angle of dip, in the same way as the spacing of topographic contours gives the angle of slope of the land surface, we can use structure contour maps to calculate angles of dip.

The block diagrams in Fig. 17 illustrate an inclined layer of rock with structure contours and direction of dip indicated. A map of the top surface of the diagram is shown in **C**. Using the map scale the horizontal distance **a − b** (in the direction of dip) can be measured, i.e. 120 m. Between **a** and **b** the top surface of the rock layer drops from a height of 200 to 100 m (see diagram **B**, **b − c**) so that the inclination of the surface can be expressed as a vertical drop of 100 m in a horizontal distance of 120 m, i.e. a gradient of 1 in 1.2.

Alternatively the *angle* of dip can be calculated: as **a–b–c** (diagrams **B** and **D**) is a right-angled triangle it follows that the tangent of the angle of dip = **bc/ab**, i.e. 100/120 = 0.83; therefore the angle of dip is about 40°.

If the inclination of the layer surface in Fig. 17 was measured in any direction other than the direction of dip, the angle of dip would be less than 40°. Thus in an oblique section, i.e. Fig. 18, the **apparent dip** is given by tan **db^cd = bc/bd**, i.e. 100/225 = 0.444, so it is 24°. It is also obvious from Figs. 17 and 18 that in a section parallel to the direction of strike, the dip would be 0°, so that oblique sections at different angles to the strike can give a range of apparent dips, in this case from 0° to 40° (the maximum and therefore the **true** angle of **dip**).

FIG. 17 Calculation of true dip using structure contours

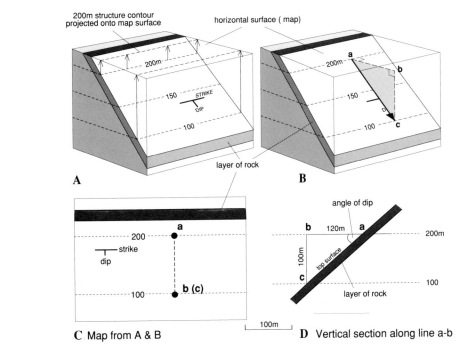

A

B

C Map from A & B

D Vertical section along line a-b

FIG. 18 Apparent dip

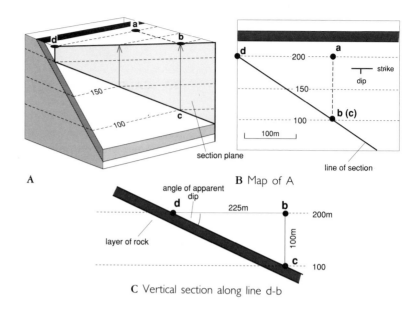

A

B Map of A

C Vertical section along line d-b

3.6 CURVI-PLANAR GEOLOGICAL SURFACES

In the preceding discussion we have considered only geological surfaces that are flat, though inclined, planes. Consequently the structure contours have been straight, parallel and equally spaced. However, natural geological surfaces are rarely so regular in shape. They are commonly curvi-planar and therefore structure contours are curved lines with variable spacing, as are topographic contours. Figure 19 illustrates two such curved surfaces and the patterns of structure contours derived from them. Note in diagrams **A**, **B** and **C** that the strike is constant but the dip variable hence the structure contours on the map (**C**) are straight but unequally spaced. In diagrams **D**, **E** and **F** the dip is nearly constant but the strike variable and therefore the structure contours are equally spaced but curved. Should *both* dip and strike vary, the pattern of structure contours would evidently be more complex. In some places in this manual straight structure contours are employed for the sake of simplicity but it must be realised that in many natural situations structure contours will not be straight, parallel, or equally spaced.

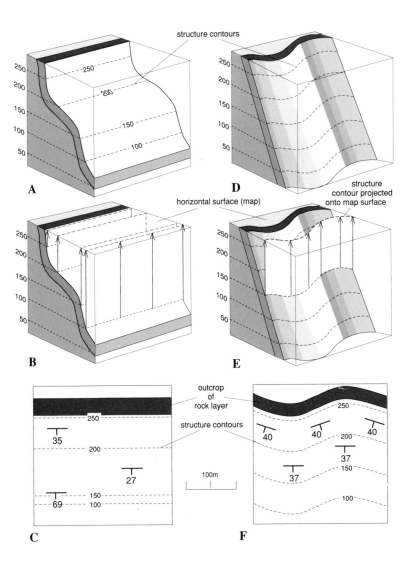

FIG. 19 Curvi-planar surfaces and structure contours

3.7 EXERCISES

The following four exercises relate to the foregoing discussion on the construction of structure contours, determining the attitudes and shapes of geological surfaces, and section construction. You should, in each case, carry out the instructions given and solve the problems posed. So that you can check your analyses and interpretations, solutions to the exercises are given in Chapter 15, at the end of the manual. You may find it convenient to undertake your analyses on an overlay of tracing paper or a photocopy of the maps.

3.7.1: Map A

FIG. 20 Exercise 3.7.1

The differently ornamented areas on the map represent the outcrops of different rock types. Planar contacts between the different rock types have different

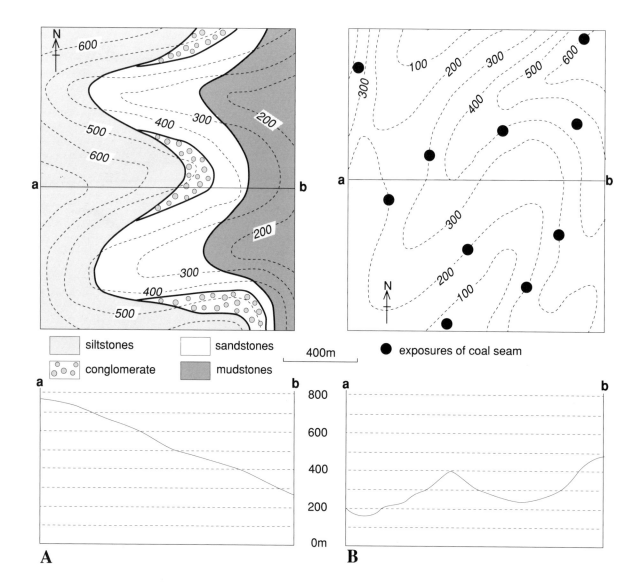

siltstones   sandstones   400m   ● exposures of coal seam

conglomerate   mudstones

attitudes. Wherever possible construct structure contours for the geological contacts shown and in each case determine their dip and strike. Using the topographic profile drawn along the line **a–b**, draw an accurate cross-section making sure you extend the traces of the geological surfaces both above, and below, ground level. *Remember that for any geological surface the more points of like height you can locate, the more accurately you can define the trend of a structure contour. In order to assess the attitude of those surfaces for which you cannot draw structure contours you will need to compare their outcrop shapes with those where the attitude is known.* Determine the vertical sequence of rock types in the map area.

3.7.1: Map B

A thin bed of coal is exposed at the points shown on the map by black circles. By constructing structure contours, determine the attitude of the bed and draw in its complete outcrop paying due attention to the effects of topography. Draw a cross-section using the profile provided and shade in, on the map, the ground lying above the bed. Determine the dip and strike of the bed. *You will need to reverse the procedure outlined in Figs 7 and 8, i.e. construct structure contours first by*

FIG. 21 Exercise 3.7.2

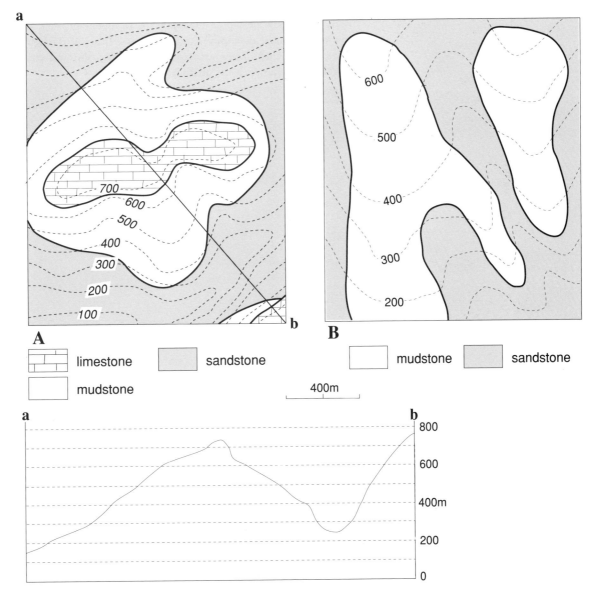

A

| | limestone | | sandstone |
| | mudstone |

B

| | mudstone | | sandstone |

400m

determining positions where, as well as already shown, structure and topographic contours of like value intersect, i.e. where the coal bed is at ground level. The full outcrop of the coal can then be drawn in by joining up these positions while giving careful consideration to the effects of topography on outcrop trend. Remember that wherever structure and topographic contours of the same value cross, the bed of coal will be at ground level (Fig. 8).

3.7.2: Map A

Using structure contours and the relationship of outcrop shape to topography, determine the attitudes of the geological contacts depicted on the map. In the case of the lower mud/sandstone contact it is apparently possible to construct two sets of structure contours; which is the more acceptable, and why? Draw an accurate cross-section along the line a–b, making sure you extend the contacts both above and below the ground surface. What are the shapes of the two surfaces and what is the vertical sequence of rock types in the area?

3.7.2: Map B

By using the relationship of outcrop shape to topography and by constructing structure contours, determine the attitude and shape of the geological contact shown on the map. Draw accurately on the map dip and strike symbols to indicate the variation in its attitude.

OUTCROP SHAPE

4.1 OUTCROP SHAPE, ANGLE OF DIP AND TOPOGRAPHY

To allow accurate and thorough analysis of geological maps it is very important to gain an understanding of the factors that control outcrop shape. Evidently a primary control will be the shape of the geological surfaces involved, i.e. whether they are flat or curvi-planar, etc. In order to illustrate the other contributing factors we will consider a layer (**bed**) of rock of uniform thickness, bounded by flat, parallel surfaces, and see how the shape of its outcrop varies with changes in topography and amount of dip.

The maps **A** to **G** in Fig. 22 show variations in the outcrop shape of a bed of rock 70 m thick that are attributable to changing angles of dip (the topography is the same). In each map north is towards the top of the page.

In A: A horizontal bed (angle of dip = 0°) shows characteristic near-parallelism of the outcrops of it's upper and lower contacts with the topographic contours. The outcrop shape is in places concentric (e.g. on the top of the hill) and is dictated by the shape of the land surface, i.e. the trend of the topographic contours.

In B: A shallow dip (14°) to the west produces a highly sinuous outcrop pattern and outcrops 'vee' in the valley in the **direction of dip**.

FIG. 22 Relationships between topography, amount of dip and outcrop shape

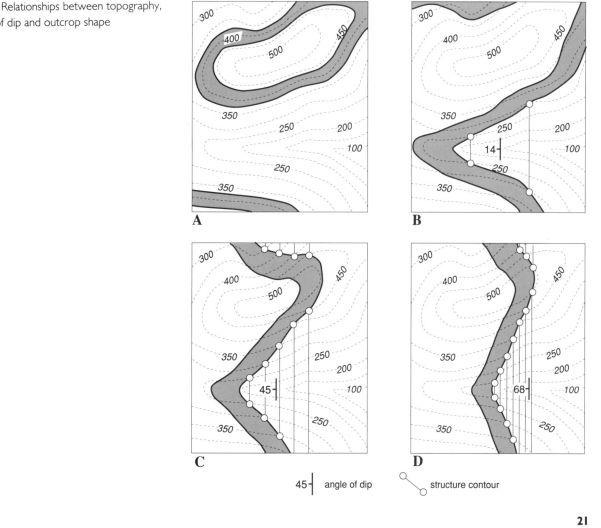

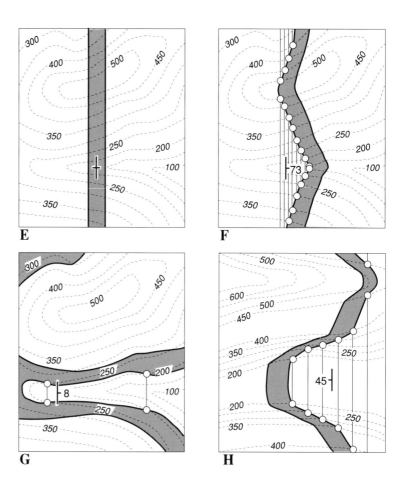

In C: A moderate dip (45°) to the west reduces the sinuosity and now the overall trend of the outcrop is in its strike direction.

In D: A steep westerly dip of 68° further reduces the sinuosity and more clearly the overall trend reflects the strike direction.

Note that with the exception of **A**, 'veeing' in the valley and over the ridge indicates the direction of dip and in **C** and **D**, the general trend of the outcrop across the map indicates the direction of strike. Note also the decrease in the spacing of the structure contours with increasing dip.

In maps **B**, **C** and **D** the bed of rock dips towards the west. In **E** it is vertical and therefore its outcrop is straight and is unaffected by topography. Clearly it trends in the direction of strike. In **F** it dips towards the east as is indicated by the structure contours and the veeing of the outcrop in the valley. In **G** the structure contours show a dip to the east at 8° but note that the vee in the valley apparently indicates a dip to the west. This arises because the slope of the valley floor is in the direction of dip at a *steeper* angle than the dip of the bed. This situation is not uncommon and should always be borne in mind, particularly when analysing maps where the dips of planar geological features are low and/or topography is rugged.

All of the maps in Fig. 22 serve to show how outcrop shape is related to direction and amount of dip, and to broad topographic features (valleys and ridges). Note, however, in map **H** that changes in the slope of the land surface also have a marked affect; the trend and width of the outcrop changes abruptly with changes in topographic slope from steep to moderate to flat. Where the outcrop crosses relatively flat ground in the floor of the valley, notice that its trend is in the direction of strike; note also that outcrops are wider where the topographic slope is low and narrower where this is steep.

Given that in each map in Fig. 22 the **true thickness** of the layer of rock is 70 m it is obvious in all but map **E** that the orthogonal width of the bed as measured on either the map surface or the land surface is greater or less than 70 m, i.e. it is an **apparent thickness**. Only where the land or map surfaces cut the layer at right angles to its contacts is the true thickness given, e.g. in diagram **E** where, as the bed is vertical, the map surface must be at right angles to the contacts of the bed.

4.2 TRUE AND APPARENT THICKNESS

Figure 23 illustrates the relationships between true and apparent thicknesses. On the map (diagram **A**) the orthogonal widths of the bed measured on the map (a′ to a‴) vary, being 50, 110, and 90 m respectively, whilst in the cross-section (diagram **B**) the bed thickness measured along the ground surface at **w** is 115 m. These are all apparent thicknesses. The true thickness (70 m) is measured in the plane at right angles to the strike, i.e. that containing the direction of dip (**t** in diagram **B**). Clearly we can measure true or apparent thicknesses by constructing cross-sections but we can also use trigonometry. The trigonometric relationships between true and apparent thickness are shown in diagrams **C**, **D** and **E**, which summarise methods used for calculating true thicknesses in the field and in the analysis of maps.

In the field, if we can measure the dip, the width of outcrop along the ground surface in the direction of dip, and the angle of slope of the topography, we can determine the true thickness of a layer. Diagram **C** of Fig. 23 shows the geometric basis for the trigonometric solution. The angle of dip is given by

FIG. 23 True and apparent thickness

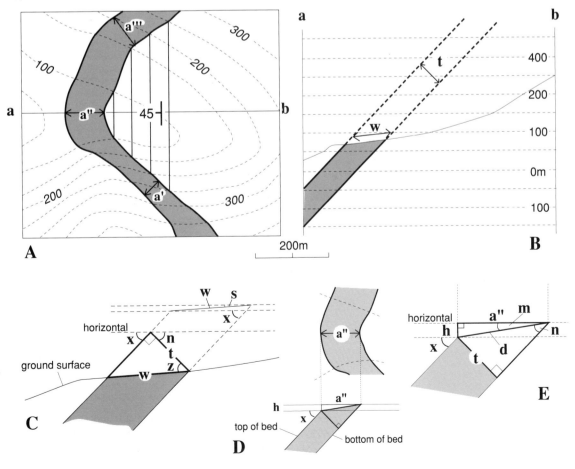

angle x, w is the width of outcrop along the ground surface, and s is its angle of slope. We need to determine the value of angle z in the right-angled triangle in order to calculate t. Angle z is equal to angle $n + s$, s we have measured and $n = 180 - (90 + x)$; therefore $z = (180 - (90 + x)) + s$. In the right-angled triangle $\cos z = t / w$ and therefore $t = w \cos z$. By substitution, $t = w \cos [\{180 - (90 + x)\} + s]$.

This calculation has to be modified when we wish to determine true thickness from a map. We need to know the dip (x), the difference in height of the outcrops of the top and bottom surfaces of the layer (h), and the width of the outcrop of the layer on the map (in this example a'', diagrams **A** and **D**). Diagram **E** is an enlargement of **D** and shows the two right-angled triangles we use in the calculation. In order to determine t we need to know the value of angle n and the distance d. We know the values of h and a'' and can therefore calculate m because $\tan m = h / a''$. Angles $m + n = x$ and therefore $n = x - m$. Distance d is given by $d = t / \cos n$. The true thickness is now given by the relationship $\sin n = t / d$ and therefore $t = d \sin n$.

4.3 CURVI-PLANAR SURFACES

Our use of the relationships between outcrop shape, angle of dip and topography to establish directions of dip and strike and approximate angles of dip, as outlined above, has to be modified when dealing with curvi-planar geological surfaces, for example that in Exercise 3.7.2B. In Fig. 24, map **A**, the outcrop shapes relative to topography suggest a dip towards the west to south-west at a moderate angle. Construction of structure contours (diagram **B**) confirms a south-westerly dip but shows a considerable variation in the angles of dip which are not readily apparent from the outcrop shapes; the surfaces are curvi-planar. Diagram **C** shows values for the dip of the two surfaces derived from analysis of the spacing of the structure contours in **B**. Thus outcrop shape, though an extremely important indicator of dip, can only give us general indications of directions and amounts of dip; in the absence of measurements of dip and strike, more detailed analysis is required to allow accurate determinations.

4.4 EXERCISES

4.4.1

In Fig. 24, map **D**, as in **A**, a superficial appraisal might suggest that the geological surfaces **s** and **r** dip towards the west in the valley and the south-west over the ridge, both at moderate to steep angles, **r** perhaps more steeply than **s**. Note, however, that whilst this is true for outcrops across the valley, in the north the outcrop of **r** crosses topographic contours with little deviation in trend, suggesting a steep dip. To check these observations construct structure contours for the two surfaces. *In analysing the map remember, wherever possible, to locate more than two points of the same height on any given surface before you decide the trend of structure contours. Check your analysis against the solution given in Chapter 15.*

4.4.2.

Further complications in interpreting outcrop patterns arise where geological surfaces are more strongly curvi-planar than those illustrated in Fig. 24, for example where rocks are deformed by folding (Fig. 25). Characteristic of the

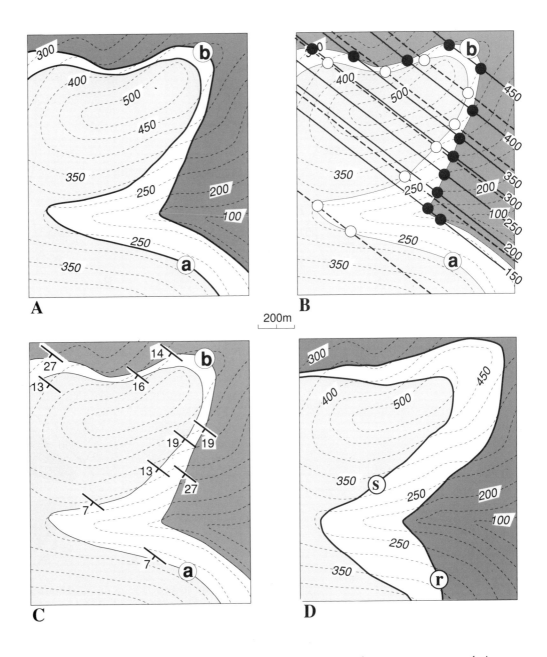

FIG. 24 Curvi-planar surfaces

outcrop patterns of folds is the combination of outcrop curvature relating to topography, i.e. over hills and ridges, and across valleys, with curvature relating to the presence of fold closures. Thus in the maps of Fig. 25 changes in outcrop trend due to topography can be readily distinguished from those due to the rapid changes in dip around fold closures.

In maps **A**, **B** and **C**, the rock layers strike north–south but dip at varying angles and directions around synformal (down) folds. Examination of outcrop shape shows that, in some places, curvature of outcrop relates to the presence of valleys and ridges whereas, in others, rapid changes in the trend of outcrops occur where the topography is uniform. In the former case 'veeing' of outcrops reveals direction of dip whilst, in the latter, it relates to the presence of fold closures where the geological surfaces become markedly curvi-planar, i.e. the amount and/or direction of dip rapidly changes. (a) For each map locate

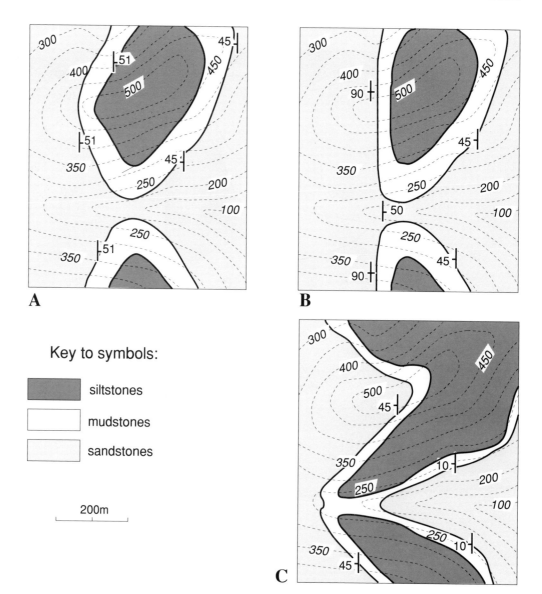

Key to symbols:

siltstones

mudstones

sandstones

200m

FIG. 25 Folds and outcrop shape

places where outcrop shape relates to dip direction and where to the presence of fold closures; (b) for maps **A** and **B** construct structure contours for the mudstone/siltstone contacts to determine the shapes of the folds; (c) for map **C** draw structure contours for both geological contacts and assess the shape of the fold; and (d) for each of the maps draw true scale, vertical, E–W cross-sections along the southern edges of the maps. *Note in **A** and **B** that it is possible to draw structure contours as a consistent N–S set or as a variable set trending between NE and SE. The latter are not however compatible with the shapes of the outcrops in relation to topography and would reflect extremely complex structures.*

4.4.3

The maps in Fig. 26 show structure contours (straight dashed lines), for a single folded geological surface. Locate points where structure and topographic contours of the same value intersect, and construct the outcrops of the surfaces. Indicate, on the map, where changes in outcrop trend are due to the influence of topography, and where to the presence of fold closures. In each case draw cross-sections on the topographic profiles provided to illustrate fold shape and check your analysis with the solutions given in Chapter 15.

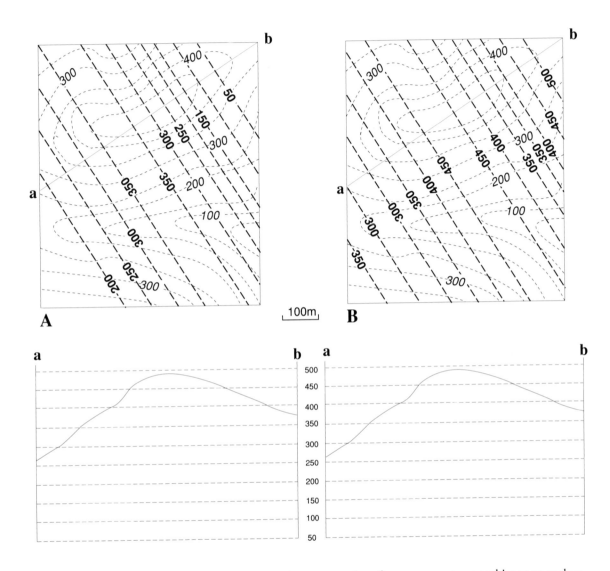

FIG. 26 Exercise 4.4.3

So far we have seen that construction of structure contours enables us to analyse the attitudes and shapes of geological surfaces—likewise, when used with care, can the relationships between outcrop shape and topography. Where topographic detail is not available, we can only rely on outcrop shape and direct readings of dip and strike as indicators of geological structure. Though lacking in precision, use of outcrop shape alone is, nevertheless, a very important method of analysis and an important consideration when mapping rocks in the field.

4.4.4

The map in Fig. 27 illustrates the geology of an area wherein the topography is indicated only by the course of streams and rivers, and by spot heights. Thus structure contours cannot be drawn nor can an accurate topographic profile be constructed. Despite this, by noting the shapes of outcrops relative to these topographic indicators we can, at least qualitatively, assess the attitudes of contacts and structures, i.e. their strike and whether they have vertical, steep, moderate, shallow or sub-horizontal dips. We can therefore gain an understanding of the structure of the area. Determine the directions and relative amounts of dip of the contacts and structures (i.e. vertical, steep, moderate, shallow or horizontal) and mark these on the map. Given that outcrops of the

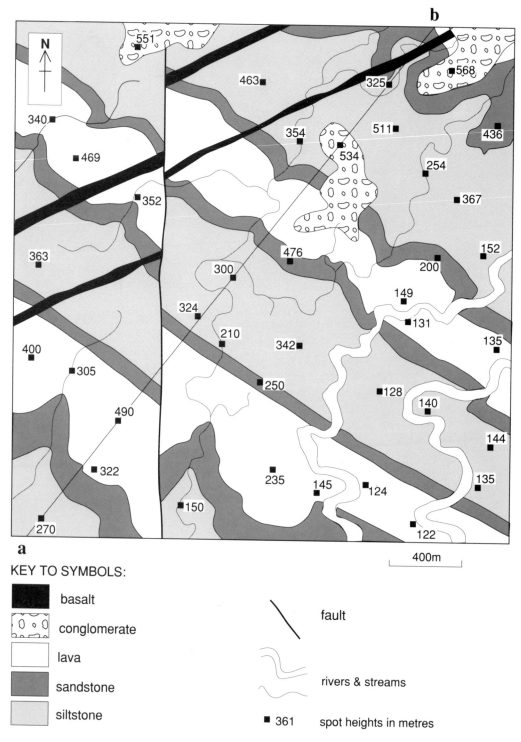

FIG. 27A Exercise 4.4.4

KEY TO SYMBOLS:

same rock type are of the same age throughout the area, draw, using the profile (Fig. 27a), a vertical cross-section along the line **a–b**. Compare your analysis with that given in Chapter 15.

FIG. 27 Exercise 4.4.4

LINEAR STRUCTURES

5.1 ORIGIN AND MEASUREMENT OF LINEAR STRUCTURES

Linear structures arise in many ways: as lines of intersection between planar elements, e.g. intersections of bedding planes and igneous contacts with each other and with unconformities; as intersections of bedding planes, igneous contacts and unconformities with faults; as a result of folding, e.g. fold crests and troughs; and, on a smaller scale, as a result of the internal strains suffered by rocks during deformation which give rise to the elongation and alignment of particles and/or crystals in rocks, in a uniform direction. Because such features are linear rather than planar, we refer to them as **linear structures** and **lineations**.

In Fig. 28, diagram **A**, a sedimentary sequence unconformably overlies an older basement complex. In such a situation there exist several planar elements such as bedding planes, dyke margins, etc., which have different attitudes. Removal of the younger sedimentary sequence, as in diagram **B**, reveals the outcrops of some of these elements, both within the basement and the sedimentary sequence, on the plane of unconformity. The outcrops comprise lines of intersection of the different geological surfaces and as such form linear features which, as will be seen later, can be important in assessing the structural geometry of an area, particularly in determining displacements on faults.

Linear structures can also arise in other ways. In situations where rocks are deformed, layered sequences of sediments frequently fold, as in Fig. 29, diagram **A**. Such folds often have linear trends over distances of centimetres to tens of kilometres and thus they constitute linear structures. Similarly, where faults cut through earlier planar structures, intersections of the latter with the fault planes will be linear. For example, in Fig. 29, diagram **B**, rocks overlying the fault plane in diagram **A** have been removed to show such linear intersections.

Evidently the attitudes of linear features and structures can be measured both in the field and on maps. However, in contrast to planar elements, this is done by reference to direction and inclination, i.e. their **plunge**, not to dip and strike as with planar elements. Figure 30 illustrates how we define and measure plunge. In the diagram a linear structure is inclined at an angle to the horizontal (top surface of the block diagram). Its attitude can be recorded by reference to its direction of inclination (its **direction of plunge**), and its angle

FIG. 28 Lines of intersection between planar elements

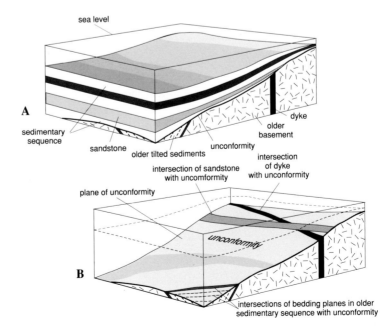

FIG. 29 Linear elements produced by folding and faulting

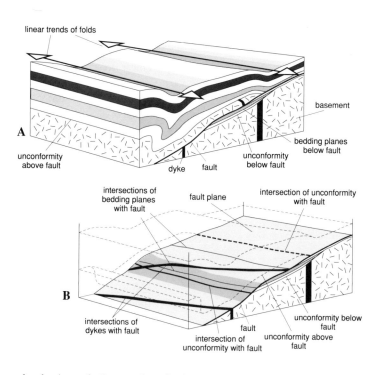

from the horizontal (its **angle of plunge**). Note that the latter is always measured in the vertical plane that contains the linear feature.

As with dip and strike, directions are referred to points of the compass, but on maps linear symbols are used (Fig. 31).

5.2 LINES OF INTERSECTION

As well as the analysis of linear structures that can be seen and measured in the field, many geological problems require construction and analysis of intersections between planar geological surfaces, such as those illustrated in Figs. 28 and 29, from the data presented on geological maps. In Fig. 32, diagram **A**, for example, conglomerates unconformably overlie and truncate a more steeply dipping, older sequence of rocks comprising siltstone and mudstone formations. Because the siltstone/mudstone contact and the unconformity are planar surfaces there must be a line of intersection between the two. How do we determine the position and attitude of this line of intersection?

The disconformable nature of the base of the conglomerate is evident on the map because (a) the conglomerates occupy the high ground and must therefore overlie the siltstone and mudstones, and (b) in three places, the base of the conglomerates cuts across the siltstone/mudstone contact (white circles in

FIG. 30 Measurement of plunge

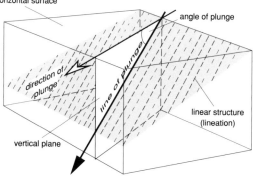

FIG. 31 Recording the attitudes of linear structures

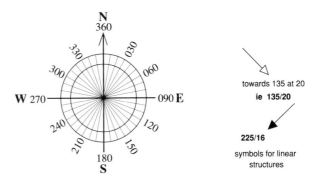

FIG. 32 Lines of intersection between planes

diagram **B**). Construction of structure contours for the geological surfaces (diagram **B**) reveals consistent dips of 40° to the West for the siltstone/mudstone contact and 17° to the south-west for the base of the conglomerates. The regular trends and spacing of the two sets of structure contours show that the two surfaces are flat but inclined planes.

By definition, the intersection of two flat planes must be a straight line

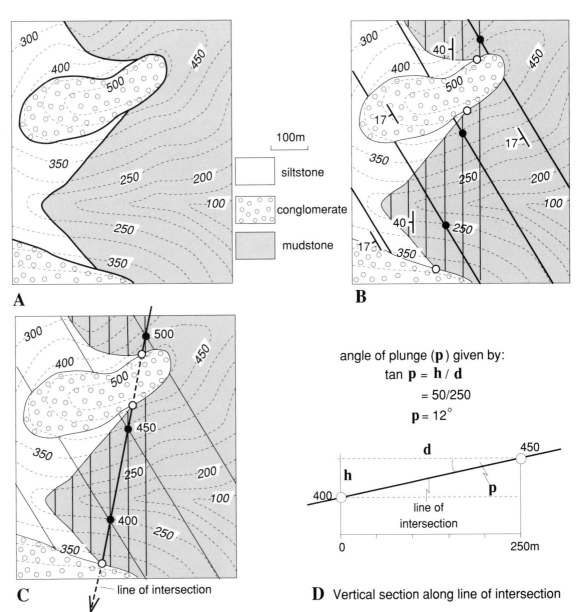

angle of plunge (p) given by:
$$\tan p = h / d$$
$$= 50/250$$
$$p = 12°$$

D Vertical section along line of intersection

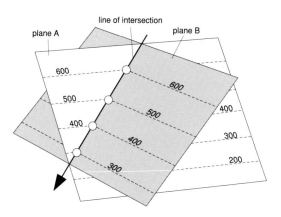

FIG. 33 Intersection of two flat, inclined planes

(Fig. 33) and it is apparent in the diagram that its position can be determined by noting where structure contours of the same height for each plane meet (white circles in Fig. 33). Thus in Fig. 32, diagram **B**, points of intersection of the plane of disconformity and the siltstone/mudstone contact can be located where structure contours of like value for the two surfaces cross (black circles). Evidently these lie on a straight line, as do the points of truncation observable directly from the map (white circles). All of these points define the line of intersection which, note, falls in height from over 500 m in the north, to less than 400 m in the south (diagram **C**). The precise attitude of this line can be calculated because in diagram **C** we can locate points on it at 500, 450 and 400 m. Therefore we know that in the horizontal distance between the 500 and 450 m points (and those at 450 and 400 m), as measured on the map (= 250 m), the line drops in height by 50 m. By trigonometry (diagram **D**) the angle of plunge is given by tan **p** = 50/250, i.e. the angle of plunge (**p**) = 12°.

Where planar surfaces are curved rather than flat, lines of intersection can be similarly constructed, though evidently they will not be straight lines.

5.3 EXERCISE

5.3.1

In Fig. 34, determine the shapes and attitudes of the geological contacts shown on the map and decide which is the unconformity. Locate the line of intersection between the two planar surfaces present and determine its direction and amount of plunge. Shade in those areas of the map where the limestone is underlain by sandstone.

Again check your analysis and interpretation by referring to the solution given in Chapter 15.

FIG. 34 Exercise 5.3.1

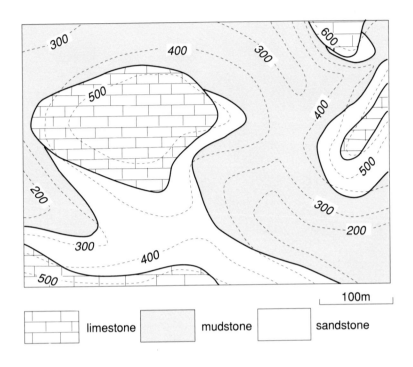

ANALYSIS OF DRILL-HOLE DATA

6.1 CALCULATION OF DIP AND STRIKE FROM DRILL-HOLE DATA

In addition to observations and measurements made at rock exposures at ground surface, geologists make use of data recovered from drill-holes and mine shafts. When analysing geological maps such information can be used to determine the attitudes of planar and linear elements lying below ground surface and is extremely useful in establishing the three-dimensional geometry of geological structures.

In the block diagram, Fig. 35 diagram **A**, vertical drill-holes **a**, **b** and **c**, pass through a coal-bearing formation (dark grey) at different depths below ground surface. The map **B** shows the location of the drill-holes. Down-hole logs for each give the following data:

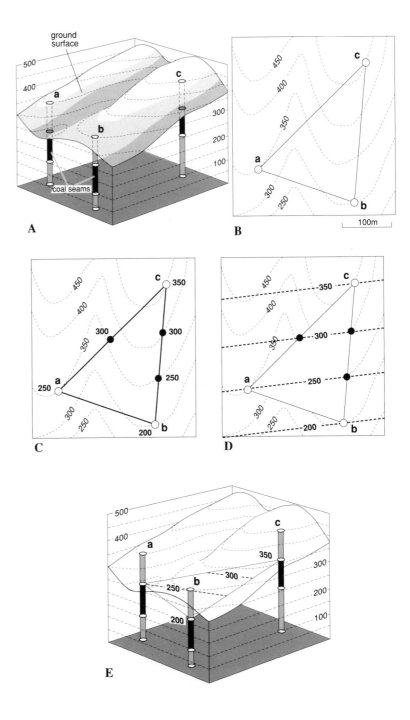

FIG. 35 Drill-hole data

	Height of ground surface	Depth to top of coal formation
a	350 m	100 m
b	300 m	100 m
c	450 m	100 m

From this information we can calculate that the top of the coal formation in **a** lies at 250 m above sea level (350 − 100 m = 250 m), in **b** at 200 m, and in **c** at 350 m. Assuming that the top surface is a flat though inclined plane, it follows that between **a** and **b** it falls from 250 to 200 m, between **b** and **c** it rises from 200 to 350 m, and between **c** and **a** it falls from 350 to 250 m (diagram **C**).

By dividing, on the map, the distance **a** to **c** in half, we can locate where the surface lies at 300 m. Likewise, by dividing **b** to **c** into three equal parts, we can locate points at 250 and 300 m (black circles in diagram **C**). This procedure gives us an array of points of known height on the surface (diagram **C**). By connecting points of like height, we can construct structure contours (diagram **D**) and therefore determine the dip and strike of the surface. Remember, however, that such a construction assumes the surface to be flat. We could be more sure of this if there were more than three drill-holes. Diagram **E** shows how this construction relates to the block diagram; the shaded, triangular area in the diagram forms that part of the top surface of the coal formation deduced by our analysis.

6.2 THREE-POINT PROBLEMS AT SURFACE

From the construction outlined above it is evident that for any flat but inclined surface, if we know the heights of three or more points on it, we can determine its attitude. This is as true for data at ground surface as it is for those recovered from drill-holes. Figure 36, diagram **A**, is a geological map showing exposures (**a** to **f**) of mudstones (dark grey), conglomerates (black circles) and siltstones (light grey) for which we only know the trends and heights of contacts. By using the 'three-point technique' explained above, we can, from this information alone, determine the likely attitudes of the contacts and construct a full geological map.

In diagram **B**, for each contact, triangles are constructed so that we can calculate intermediate heights and from these draw in appropriate structure contours (diagram **C**). Evidently the construction for the siltstone/conglomerate contact is more reliable than that for the mudstone/conglomerate because there are six points of known height rather than four. Given this, we can however measure dip and strike. Also, by noting where the structure contours cross topographic contours of like value, we can predict where the contacts are near surface (white circles in diagram **D**) and therefore we can complete the geological map.

In these examples we are assuming that the geological surfaces are relatively flat though inclined planes, i.e. their dip and strike are consistent. Whether or not this is the case can be judged if, for example, more than three reference points are available; trends of contacts are appropriate (as they are in Fig. 36); or the distribution of other exposures is compatible.

FIG. 36 Three-point solutions

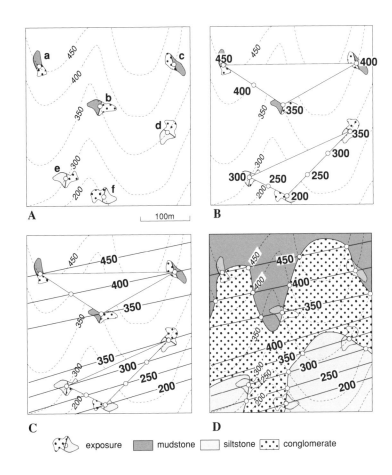

exposure ▢ mudstone ▢ siltstone ▢ conglomerate

6.3 EXERCISE

6.3.1

In Fig. 37 vertical drill-holes at localities **A**, **B**, **C** and **D** pass through the following formation boundaries at the depths given, in metres below ground surface:

	A	B	C	D
Base of sandstones	50	50	150	50
Top of gold-bearing conglomerate	300	100	350	Absent
Bottom of gold-bearing conglomerate	350	150	400	Absent

Note in drill-hole **D** that the conglomerate is not present. Assuming that these surfaces are flat planes, construct structure contours and draw in their outcrops. In advising a mining company who wish to explore for gold, which area of the map would you predict is underlain by the conglomerate? Shade this area on the map and draw an N–S cross-section to illustrate the structure. In tackling this exercise you will need to determine the positions and attitudes of the three geological surfaces from the drill-hole data; locate any lines of intersection that may be present; and determine the structure of the area.

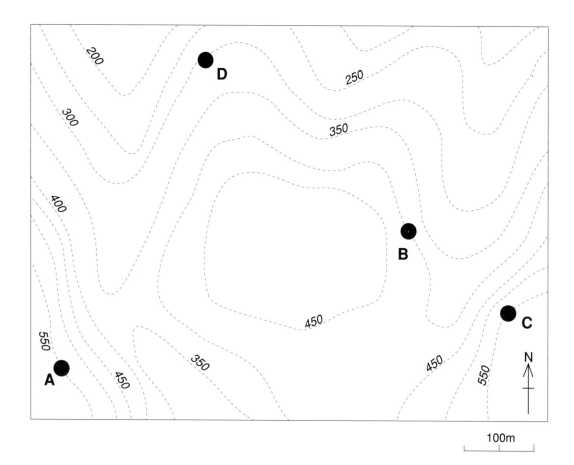

FIG. 37 Exercise 6.3.1

ISOPACHYTES (THICKNESS CONTOURS)

7.1 CALCULATION OF THICKNESS VARIATIONS

Thus far we have seen how the three-dimensional shapes of geological surfaces can be analysed using measurements of dip and strike, structure contours and/or outcrop shapes in relation to topography. Similar techniques can also be utilised to analyse variations in the thicknesses and volumes of bodies of rock. For example, some sedimentary formations or beds may vary in thickness in relation to the channel systems in which they were deposited. Study of thickness variations allows us to reconstruct these systems and therefore appreciate their three-dimensional form and, perhaps, their origin. Assessment of shape, and volume calculations, are also important in establishing the economic potential of ore bodies, whether these be sedimentary, igneous or volcanic in origin.

Where on maps, with or without drill-hole data, we are able to construct structure contours for the top and bottom surfaces of bodies of rock, we can also construct thickness contours (**isopachytes** or **isopachs**). In simple situations on maps, where structure contours for top and bottom surfaces cross, vertical thickness can be readily calculated by deducting the lower value for height from the higher. Thus in Fig. 38, diagram **A**, structure contours for upper (**a**) and lower (**b**) curvi-planar surfaces bounding a body of rock intersect in several places. These are shown in diagram **B** as black squares against which values for vertical thickness are given. Contouring of these thickness values gives lines of equal thickness, i.e. isopachytes (thick dashed lines). Note that the body of rock is roughly lensoidal in shape (block diagram **C**). In this example, however, we have only considered variation in vertical thickness, not true thickness. Should it be necessary to determine the latter we can calculate this either by constructing several cross-sections and physically measuring variations, or by trigonometry. Figure 39, diagram **A**, is a vertical section drawn where the structure contours for the top and bottom surfaces of a body of rock intersect; the vertical thickness is 200 m. We can calculate the true thickness at this intersection because we can determine the angles of dip, or apparent dip, from the map and we know the vertical thickness. The dip (**x** in diagram **B**) is given by the spacing of the structure contours. The true thickness (**t**) would be measured at right angles to the most evenly dipping surface (in this case the

FIG. 38 Construction of isopachytes

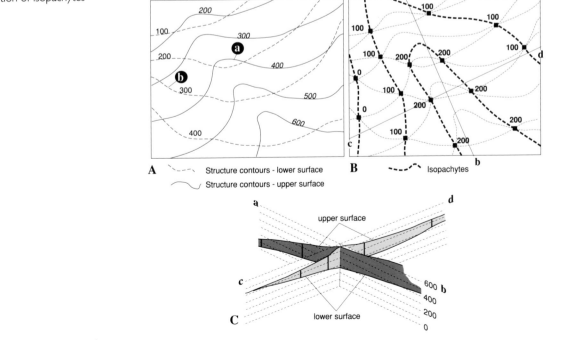

A   - - - Structure contours - lower surface
         Structure contours - upper surface

B   - - - Isopachytes

FIG. 39 Calculation of true thickness

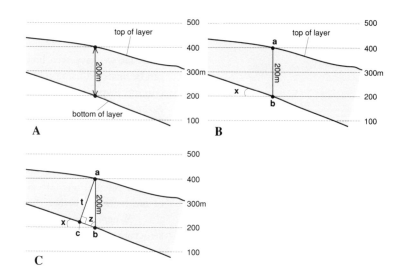

lower – diagram **C**). The vertical thickness, true thickness and dip are related by the right-angled triangle **abc**. Angle **z** (the angle between the surface and the line of vertical thickness) can be calculated: it is equal to $90° - $ **x**, and it follows that sin **z** = **t** / **ab**. Thus sin $90° - $ **x** = **t** / **ab** and from this, in this example, **t** = sin $(90° - $ **x**$)$ / 200.

7.2 EXERCISES

FIG. 40 Exercise 7.2.1.

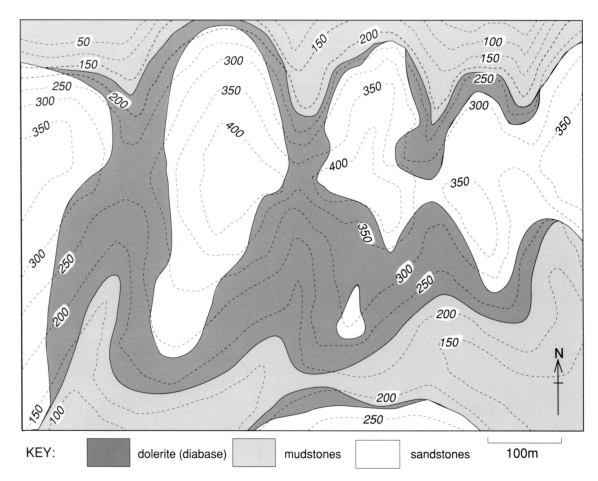

KEY: ▒ dolerite (diabase) ░ mudstones □ sandstones   100m

7.2.1

In Fig. 40 intruded igneous material has crystallised as dolerite (diabase) along the contact between sandstones and mudstones. Using the relationships between outcrop shape and topography, determine the general attitudes and shapes of the contacts between the different rock types and on this basis construct structure contours for the upper and lower surfaces of the dolerite. By noting where structure contours for the two surfaces cross, determine the variation in vertical thickness of the dolerite over the whole map area. Construct isopachytes for the dolerite by contouring these thickness values and construct N–S and E–W sections to illustrate its shape.

FIG. 41 Exercise 7.2.2.

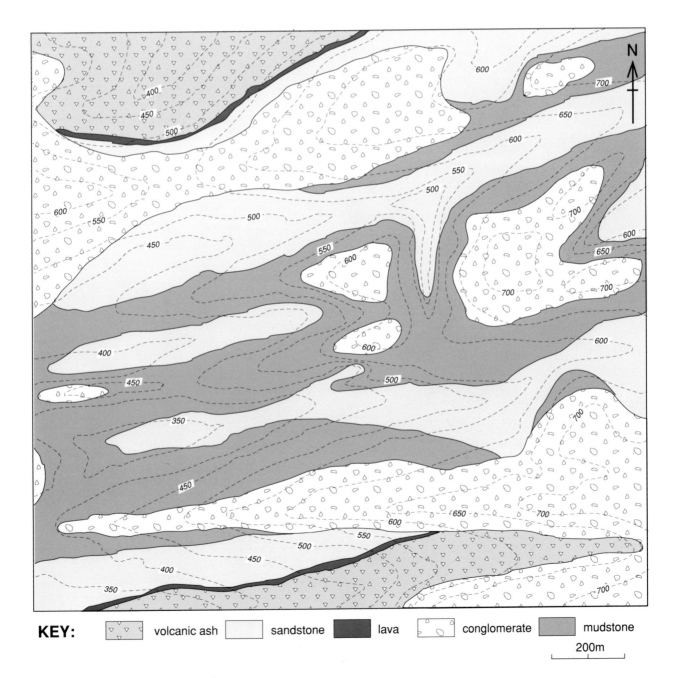

KEY: ▽▽▽ volcanic ash   sandstone   ■ lava   conglomerate   mudstone

200m

7.2.2

The mudstone in the map area shown in Fig. 41 contains barium at high enough concentrations to make extraction economically viable if reserves are sufficient. A mining company would therefore need to know the extent and thickness of the formation and particularly where it is thickest. Analyse the structure of the map area to determine the lateral extent of the mudstone and determine the variation in its thickness both before and after erosion of the present topography.

FAULTS AND FAULT MOVEMENTS

8.1 ASSESSMENT OF MOVEMENTS ACROSS FAULTS

Faults are fractures of the Earth's crust that arise when directed stresses, mostly caused by plate movements, exceed the strength of rock materials. When movement along faults is sudden, the shock waves produced are responsible for the generation of earthquakes. Though individual movements are small, displacements being measured in centimetres to tens of metres, over long periods of time (millions of years) successive small movements can give rise to very large displacements of hundreds of metres to hundreds of kilometres.

Accurate assessment of the movements across faults is important not only in situations where rock bodies of economic importance are displaced, but also when we wish to understand how different types of fault are formed, and when we wish to investigate the stress systems that have operated during the geological evolution of an area. From the data presented on a geological map we can often readily assess the relative movements (or **shift**) across a fault, but determination of the actual amount and direction of **slip** frequently demands detailed analysis. For example, in Fig. 42 (diagrams **A** and **B**) a fault has displaced a geological surface such that the left-hand side appears to have moved down relative to the right-hand side—the **downthrow** of the geological surface across the fault is to the left. We can measure the amount of offset by reference to structure contours drawn on both the geological surface and the fault plane. Thus in diagram **B** the surface has been shifted horizontally by distance **s**, and down the dip of the fault plane by distance **d**. However, whilst such information accurately tells us the amounts of shift across the fault, it does not allow us to determine the actual direction and amount of slip.

Figure 42, diagrams **C**, **D**, **E** and **F** show that the same amounts of shift can be achieved by slip (arrows) respectively: down the dip of the fault plane, by slip along the strike of the fault plane, obliquely along the fault, or by more complex movement. In each case the downthrown side is the same, as are the amounts and directions of shift. Evidently to determine directions of slip rather than shift we need more information.

Where fault planes are exposed in the field they sometimes show striations, step structures and oriented mineral fibres which give us direct evidence for the direction of slip. However, in many cases, the only indication of the presence of a fault is an abrupt change in the disposition and/or attitude of rocks across unexposed ground. In these cases analysis of geological maps, rather than direct observation, may allow us to assess both the amounts and directions of slip across faults as well as the shift.

FIG. 42 Displacement across a fault

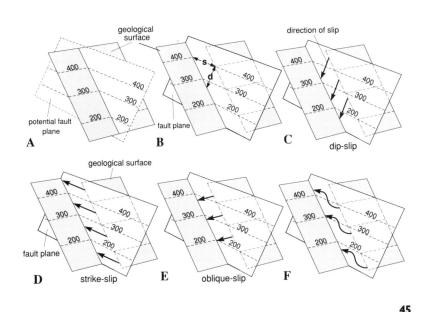

8.2 EXERCISE

8.2.1

In Fig. 43 the shaded outcrop contains iron ore and is displaced by fault **f**. It is intended to mine the ore and a mining company wishes to know the full underground extent of the iron-bearing formation. In tackling such a problem a geologist would not only need to determine the attitude of the iron formation but also the effects of the fault. To achieve this it is necessary to construct structure contours for the fault and the upper and lower boundaries of the iron formation. To assess the affects of the fault you will need, on both sides of the fault, to construct the intersections (see Figs 32 and 33) of the iron formation with the fault plane. In doing this you are making a map of the outcrop of the iron formation *on* the fault plane. When you have completed these

FIG. 43 Exercise 8.2.1.

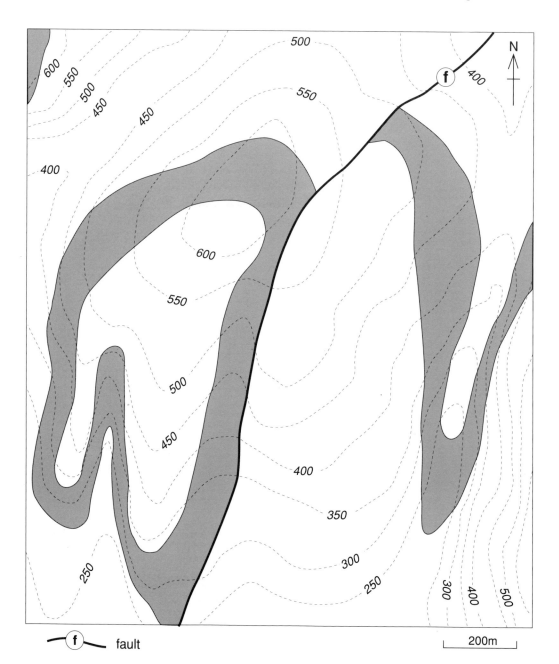

f ⟍ fault

200m

constructions, determine the along strike and down dip shift across the fault and shade in the area of the map that is underlain by the iron formation. Assuming unidirectional movement on the fault during its formation, can you determine the direction and amount of slip on the fault?

Having attempted to analyse the map, check your methods and results against the explanation that follows.

It is clear from the relationships between outcrop shape and topography that the fault plane dips towards the east-south-east at a moderate angle and the iron formation, on both sides of the fault, dips at a shallow angle to the south-south-west. These attitudes are confirmed in diagram **A** of Fig. 44 which indicates the positions of structure contours for the fault and iron formation that can be deduced from the intersections of the outcrops with the topographic contours. These demonstrate that the fault dips consistently at 48° to the south-east and the upper and lower boundaries of the iron formation are parallel, flat but inclined planes, dipping at 19° to the south-south-west. In diagram **B**, the lines of intersection of the lower boundary of the iron formation with both sides of the fault plane are constructed by noting positions (small circles) where structure contours of like height intersect (cf. Fig. 32). In diagram **C** the intersections of the upper surface have also been constructed so that the outcrops of the iron formation on the fault plane can be mapped in (light grey bands). Movement on the fault has shifted the iron formation along the strike of the fault by 530 m and along the dip of the fault by 154 m as measured ON the fault plane (**s** and **d** in diagram **D**); note that the shift along

FIG. 44 Analysis of the map in Fig. 43

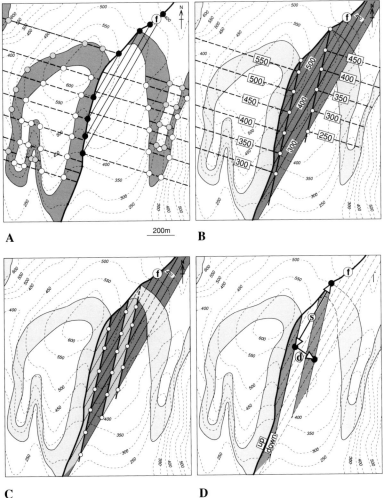

FIG. 45 Area of Fig. 43 underlain by iron formation

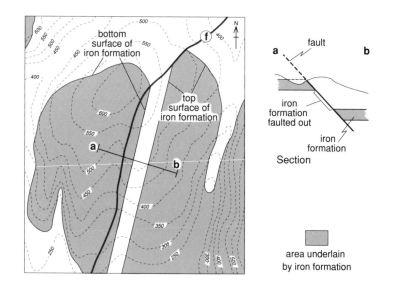

the direction of dip, as measured on the map surface, is 140 m but because the fault dips at 48° SE, that *on the fault plane* is 154 m.

The shift on the fault is like that illustrated in Fig. 42 and could be caused by movement along the strike of the fault, along its dip direction, or obliquely. It is important to realise that when only single planar surfaces, or sets of parallel planar surfaces, are displaced across faults, we cannot determine the direction and amount of *slip* on the fault, we can only measure the *shift*. Though we can say that the eastern side of the fault in Fig. 44, diagram **D**, has moved down relative to the western, i.e. the downthrow is to the south-east, this does not imply actual slip down the dip of the fault.

The areas underlain by the iron formation are bounded, to the west of the fault, by both the surface outcrop of the lower boundary of the iron (it dips to the south-south-west) and the outcrop of this boundary on the fault plane (Fig. 45). To the east it is bounded by the surface outcrop of the same boundary and the outcrop of the *upper* boundary on the fault plane (Fig. 45).

The analysis given above shows that whilst there may not be enough information to determine the direction and amount of slip on a fault, we can nevertheless assess the effect of faulting on the disposition of bodies of rock. In situations where all we need to know is the underground extent of a particular formation or structure, it is often enough to know the shift across a fault to be sure of our interpretation. However, in analysing the type of deformation that is responsible for faulting, that is whether dip-, strike- or oblique-slip is involved, we need more information. In addition, when we wish to assess the geometry of causative stress systems it is essential to determine amounts and directions of slip rather than the shift because it is these that allow us to deduce directions of compression and extension.

It is very important to understand the problems involved in distinguishing between slip direction and direction of shift (offset). As a further example, Fig. 46 illustrates displacement of a layer of rock across a vertical fault. In diagram **A** the shift of the black layer on the top surface of the diagram, which could represent a sub-horizontal topographic surface, suggests slip along the strike of the fault. However, on the vertical face (a sub-vertical cliff face), shift of the outcrop suggests up-and-down movement. This is a common situation and can arise in different ways. Displacement could, for example, be due to oblique-slip on the fault plane, as in diagram **B**. Equally the same outcrop pattern could arise from strike-slip movement, as in diagrams **C** and **D**.

FIG. 46 Offsets of outcrop and slip across faults

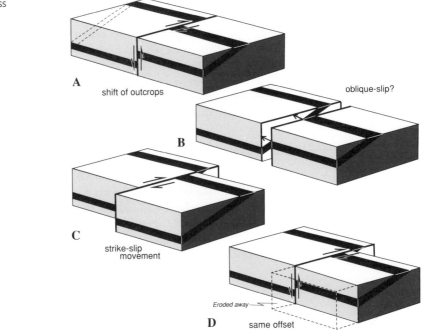

8.3 DETERMINATION OF SLIP DIRECTIONS ON FAULTS

Recognition of slip directions on faults from data recorded on geological maps relies on, amongst other criteria, recognising linear geological features that occur on either side of the fault and that were continuous before faulting. Axes of curvature of curvi-planar geological surfaces (such as folds) and intersections of flat planar surfaces of different attitudes are examples of such features. In Fig. 47, diagrams **A** and **B**, a linear structure is displaced across a fault whose slip direction can be determined by noting the relative direction of movement of points **x** and **y** across the fault plane. These are points where the linear feature intersects the fault plane which were, before movement , in the same place (i.e. at **a** in diagram **A**). If we assume a constant direction of movement during formation of the fault, the slip direction is given by the outlined arrow in diagram **B**.

It is very important to remember, however, that we are assuming a constant direction of slip during formation of the fault; this may not always be the case. Direct observations of the attitudes of indicators of slip vectors such as striations, steps and mineral fibre growths on fault planes (known as **slickensides** and **slickolites**) give the best evidence for slip directions.

As well as linear features, vertical planar structures, such as dykes, are sometimes useful in assessing slip directions. Figure 48 shows some of the

FIG. 47 Displacement of a linear feature across a fault

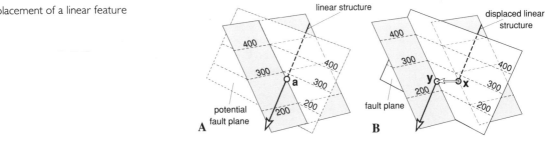

characteristics of displacements of such structures.

In diagram **A** of Fig. 48 the trend of the vertical dyke is at right angles to the strike of the fault and the slip direction on the fault is down the dip (diagram **B**). The slip direction is therefore coincident with the line of intersection of the dyke and the fault and consequently, on the map (diagram **C**), there is no offset of the dyke **a** across the fault. If however the dyke changes thickness with depth, a situation like **b** would occur. In both cases the disposition of the dykes across the fault shows that slip must have been in the dip direction of the fault; there is no horizontal shift of the dykes across the fault.

Where the angle between the strike of the dyke and that of the fault is less than 90° (diagram **D**), the direction of slip and the line of intersection of the dyke with the fault no longer coincide and therefore, on the map, the dyke is offset by an amount proportional to the amount of slip (diagrams **D** and **E**). However, from the map alone such an offset could arise from dip-, strike- or oblique-slip, as in Fig. 42. This is illustrated in diagram **F** of Fig. 48, wherein structure contours for the fault and the intersections of the dyke with the fault plane are shown together with some of the possible slip vectors (arrows). Note that the shift (or offset) of the dyke across the fault does not neccessarily imply strike-slip movement.

The foregoing discussion emphasises the need, in assessing displacements across faults, for recognising linear features that are displaced. Where intrusive rocks are involved, intersections of dykes of different attitudes or dykes with other planar surfaces provide such linear features.

In Fig. 49, diagram **A**, a fault causes offsetting of two dykes so that on the map the outcrop pattern shows a left-hand shift across the fault. As both dykes show the same sense of shift it might be assumed that slip along the fault must be along its strike. Note however that the amount of offset (**s**), is not the same. Were this so the displacement would be strike-slip. To determine the slip we need to find a linear feature and this is provided by the line of intersection of the dykes. As both dykes are vertical, their intersections with each other on either side of the fault must also be vertical. The positions of the intersections of one side of the intrusions are located at **w** and **x**, relating to the north and south sides of the fault respectively (diagram **C**). Intersection **w** lies below the fault whereas **x** lies above (the fault dips to the south-west). Because

FIG. 48 Displacement of dykes across faults

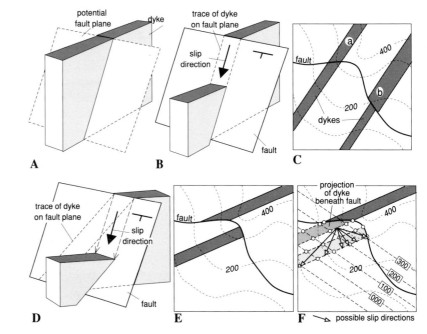

FIG. 49 Displacement of intersection across faults

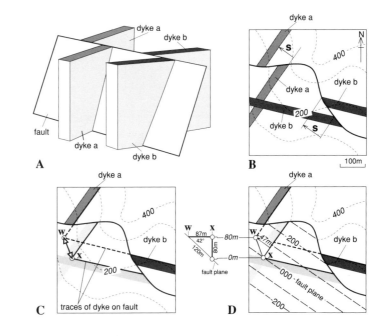

these intersections are vertical lines they must hit the fault plane directly below and above points **w** and **x** respectively. As they were continuous before faulting, the line connecting them on the fault plane gives the slip vector (arrow in diagram **C**).

Displacement on the fault involved oblique-slip with either the block to the south-west of the fault moving down relative to the North-East, or the north-eastern block moving up. Because we can calculate from the map the difference in height between the points of intersection **w** and **x** and we can determine the angle of apparent dip of the fault plane in the slip direction (diagram **D**), we can measure the amount of slip. The fault has caused a shift of about 47 m horizontally and 80 m vertically. The slip vector however trends from 338 to 158 and the amount of slip in this direction on the fault plane is 120 m (diagram **D**). Note however that in this particular case the offset of dyke **a** alone, because it is vertical and trends at right angles to the strike of the fault, shows that slip on the fault must have involved a component of horizontal movement. This is because, unlike the case in Fig. 48 (diagram **C**) the dyke outcrop is offset across the fault.

8.4 EXERCISE

8.4.1

The fault in Fig. 50 cuts and displaces the contact between a mudstone and a sandstone as well as an intrusion of basalt. Analyse the structure of the map and, assuming unidirectional movement on the fault, determine the amount and direction of slip. *In order to assess the slip on the fault you will need to determine the attitudes of the fault, the contact between the mudstones and sandstones, and the bodies of basalt. You will also need to find, and determine the attitude of, a line of intersection which, before faulting, was continuous. As in Fig. 47, locate where this hits both sides of the fault plane.*

FIG. 50 Exercise 8.4.1

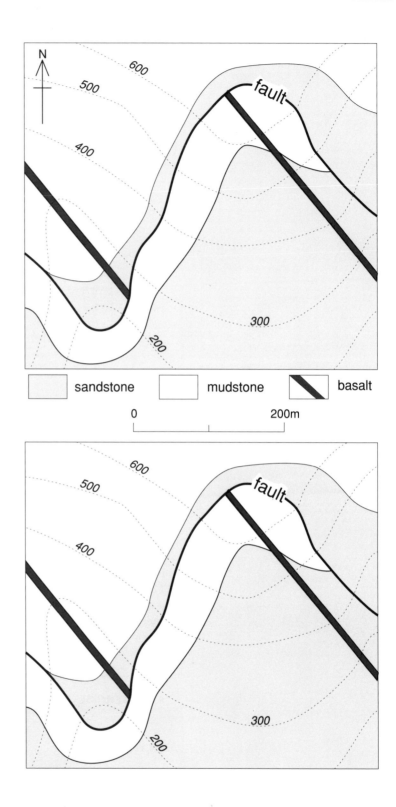

sandstone   mudstone   basalt

0   200m

TYPES OF FAULT

9.1 CLASSIFICATION OF FAULTS

Simple classifications of faults recognise an association between fault attitude and slip direction, reflecting the fact that the orientations of the stress directions responsible for crustal fracturing are generally at right angles, and parallel to, the surface of the Earth. Commonly described types of faults are illustrated in Fig. 51.

In diagram **A**, inclined layers of rock are shown in a block diagram, the top surface of which is horizontal, i.e. parallel to the surface of the Earth. In **B** the slip direction (arrows) on a fault dipping at about 60°, is *down the dip* of the fault plane; faults with these characteristics are classified as **normal faults**. In C slip is *up the dip direction* of the fault plane, which dips at about 30°—a **thrust fault**. In **D** slip is *along the strike* of a vertical fault plane, i.e. the slip direction is horizontal—a **tear fault** (also known as a **strike-slip**, or **transcurrent fault**). In **C**, were the dip of the fault plane more than 45°, the fault would be called a **reverse fault**. Note that a normal fault causes extension of the crust in a horizontal direction at right angles to its strike whilst a thrust (or a reverse) fault causes contraction in the same direction. A strike-slip fault causes both horizontal extension and contraction oblique to the strike of the fault. Faults such as these are developed on scales of a few centimetres to several kilometres, but not all faults found belong to these categories.

As slip on faults is not always in the dip or strike direction, the simple classification given above has to be modified. Figure 52 illustrates some of the variations seen in the relationships between fault attitude and movement direction. Although incomplete, this type of descriptive classification is preferable because it does not assume that stresses always act at right angles or parallel to the surface of the Earth or that, once formed, faults are not rotated.

Additional complications arise from the fact that many faults are curvi-planar rather than flat and, as we will see later, faults can develop as more complex systems than those depicted.

Figure 53 illustrates the geometry of simple curvi-planar faults which are referred to as **listric** (curved) thrust and normal faults. Characteristic of listric faults are systematic variations in the angles of dip so that in places the fault planes lie parallel or nearly parallel to the surface of the earth and consequently,

FIG. 51 Simple types of faults

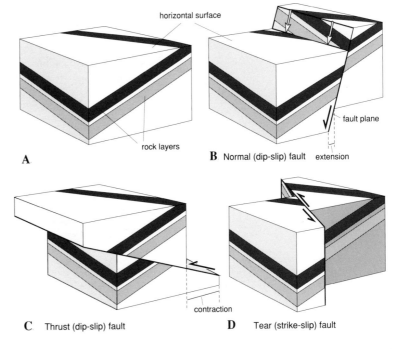

A

B Normal (dip-slip) fault   extension

C Thrust (dip-slip) fault

D Tear (strike-slip) fault

FIG. 52 Slip-dip classification of faults

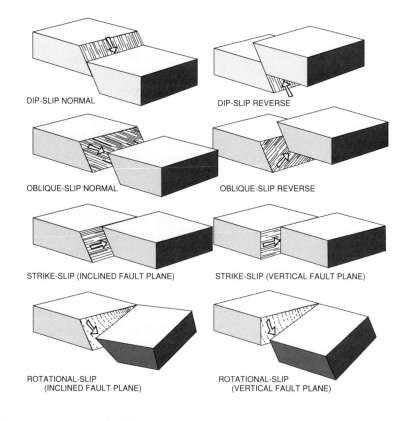

in many sequences of sedimentary rocks, parallel to the bedding. In other places they progressively increase in dip and cross-cut bedding (Fig. 53).

9.2 LISTRIC THRUST FAULTS

In Fig. 54 a listric thrust fault (due to compression) is propagated through a sequence of flat-lying sediments. It generates at first parallel to the layering along a **flat** (diagram **B**), then climbs up through the layering along a **ramp**, flattening off again in the movement direction as a further flat (diagrams **B** and **C**). Note that material lying above the fault is said to be in the **hanging-wall** of the fault whilst that below is in the **footwall** (diagram **C**). Though dips of thrust ramps are usually less than 30°, where they exceed 45° they would, in the simple classification of faults that was outlined earlier, be classified as reverse faults.

FIG. 53 Listric faults

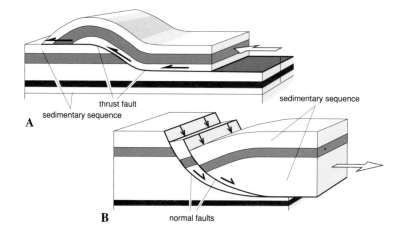

As a consequence of movement up and over the ramp, rocks above the fault become folded into an antiform/synform pair (diagrams **B** and **C**). Such folds are typical associated structures of listric thrust faults. A further characteristic of thrust faults is that slip, which is often but not always up the dip of the ramps, causes not only contraction but also thickening of the crust (i.e. above the ramp and upper flat in diagram **C**). Slip on the fault has caused displacement of the hanging-wall by distance **d** (diagram **C**).

Where large-scale thrust faults are formed, with displacements measured in kilometres, the transported rock material overlying the fault plane (in the hanging-wall) is often referred to as a **thrust nappe**. Commonly, rather than developing as single faults, thrust faults occur as complex interlinked systems known as **thrust complexes** or **zones**. Across these, total displacements may be of the order of tens of kilometres.

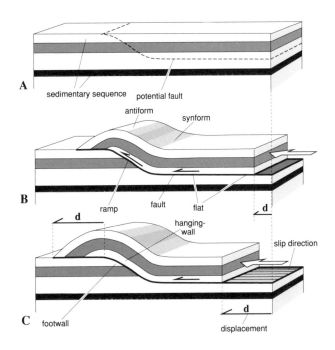

FIG. 54 Development of a thrust fault

9.3 LISTRIC NORMAL FAULTS

Listric normal faults involve crustal extension and often, but not always, slip down the dip of the fault plane. Their curvi-planar shapes involve low-angled flats and steeply dipping ramps, as do listric thrusts. As extension proceeds, the potential dilational gap shown in diagram **B** of Fig. 55 is prevented by collapse, under gravitational forces, of the rock material in the hanging-wall (diagram **C**). This either gives rise to a **roll-over antiform** adjacent to the fault plane, and/or subsidiary, **antithetic normal faults**, dipping and downthrowing back towards the main fault (diagrams **C** and **D**). Because the antithetic and main extensional fault in diagram **D** are formed during the same phase of deformation and the dip and downthrow are in opposite directions, they are referred to as **conjugate faults**. Alternatively, further extension can be accommodated by formation of one or more **synthetic faults** which develop sub-parallel to the first (diagram **E**, fault 2). However, in many cases, extensional fault systems may involve both synthetic and conjugate antithetic faults, so that fault zones may be very complex. As with other types of fault, such systems can achieve very large combined displacements and, in the case of extensional faulting, can give rise to subsidence of the crust and the formation of major sedimentary basins.

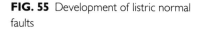

FIG. 55 Development of listric normal faults

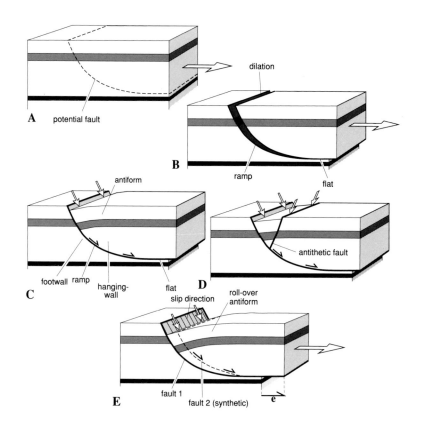

Conjugate systems of faults are not restricted to extensional faulting; they can occur in association with all other types of fault.

9.4 LINKED FAULT SYSTEMS

The occurrence of complex compressional and extensional fault systems further modifies the simple classification into normal, reverse, thrust and tear faults outlined earlier. In both compressional and extensional situations (including that of strike-slip) not only can the attitudes of fault planes be extremely variable and conjugate systems form, but also the slip directions on fault planes do not always directly relate to dip. Thus, for example (as in Fig. 52), we can find vertical fault planes across which vertical- or oblique-slip has taken place and low-angle faults that are associated with strike-slip. As an example of such possible complexities Fig. 56 illustrates a situation where rocks have failed along a linked system of listric normal faults.

The two listric normal faults **a** and **b** are linked by the vertical fault **c – d** which, as all these faults formed at the same time, is known as a **transfer fault**. It transfers displacement on fault **a** to fault **b**. The slip direction is oblique to its dip and strike and movement on it is determined by the movement sense on the controlling faults **a** and **b**. The amount of slip on the transfer fault varies along its length, as does the sense of downthrow. Between **c** and **e** the downthrow is on the left, between **e** and **d** on the right. Thus, whilst the fault is vertical and slip includes a component of horizontal displacement, it is not according to the strict definition a strike-slip fault. Because formation of the roll-over antiforms involves rotation, slip vectors are curved and therefore it is a complex **rotational fault**.

On maps, transfer faults are characterised by abrupt linking of faults with different trends and by changes in downthrow along individual faults. For example, in Fig. 57, two extensional faults trending NE–SW are linked by a

FIG. 56 Oblique-slip caused by transfer faulting

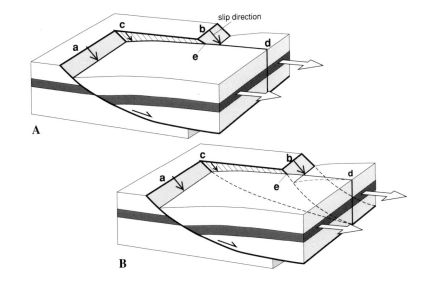

NW–SE trending transfer fault. Note that along the transfer fault the downthrow side changes where the northernmost extensional fault joins it. The increase in the dip of the bedding towards the extensional faults indicates the development of a roll-over anticline and therefore a listric shape for the faults.

Linked systems of faults are common and are associated with compressional as well as extensional faulting. For example, in Fig. 58, strike-slip faults with the same sense of displacement are linked by an extensional fault developed as a consequence of the offsetting of the major fault planes. Because of the sense of movement on the strike-slip faults and their sense of offset, extensional stresses are generated in the transfer zone and are relieved by faulting.

As movement on the main faults proceeds, development of a listric extensional fault accommodates extension of the rocks in the transfer zone. Hanging-wall collapse causes a **pull-apart basin** to form (diagrams **A** and **B**). The transfer fault linking the major faults in Fig. 58 is seen in plan view in diagram **A**. Note that the slip direction (arrow) is not down the dip of the fault plane but is in the strike direction of the major strike-slip faults.

A consequence of this type of transfer faulting is that, because of the formation of the roll-over antiform, rotational-slip must take place on the strike-slip faults within the transfer zone. This is illustrated in a section drawn along one or other of these faults (Fig. 59, diagram **B**). Note that whilst the net movement (large arrow) is horizontal, collapse of the hanging-wall of the listric extensional fault block, as a roll-over antiform, causes rotational-slip. Beyond the pivot points, however, slip on the strike-slip faults is entirely horizontal.

FIG. 57 Transfer faulting

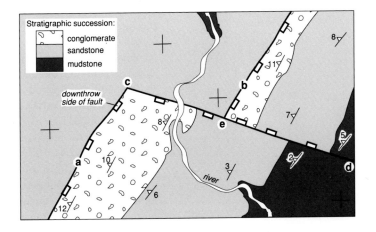

FIG. 58 Linked strike-slip faults

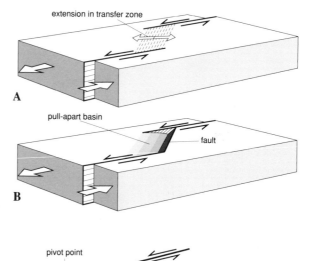

FIG. 59 Slip directions on linked strike-slip faults

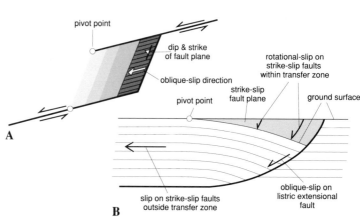

9.5 VARIATION IN SLIP ALONG FAULTS

Movement of the rocks in the transfer zone of Fig. 59 evidently involves a rotational component. Note also that the amount of slip systematically increases away from the pivot points. Variations in the directions and amounts of slip on faults can also arise because fault planes are limited in extent—displacements must eventually die away laterally as faults, however big, do not circumnavigate the world!

As an example, the faults illustrated in Fig. 60 are conjugate extensional (normal) faults that cause sagging of the crust and development of a rift valley. Many of the faults are impersistent along strike and amounts of slip therefore vary from zero to a maximum. Whether or not slip is rotational depends on the mechanism whereby rocks in the hanging-wall subside. Uniform transport in the dip direction of the faults (Fig. 60 diagram **B**) gives unidirectional but variable amounts of slip. On the other hand, rotation and stretching of the hanging-wall material (diagram **C**) produces variable amounts of rotational slip. Thus, as faults die away laterally in both directions along strike, variations in the amounts and directions of slip as illustrated in diagrams **D** and **E** can occur. Such lateral impersistence, and its consequences, are true of compressional as well as extensional faults.

Whilst this brief discussion of types of faults and fault systems does not cover all situations, it serves to illustrate that the classification of faults into normal, reverse, thrust and strike-slip faults involving only dip-slip, or strike-slip is an over-simplification. Investigation of the geometry of faults and fault movements requires careful three-dimensional analysis. It is very easy to make mistakes in interpretation unless a thorough analysis is undertaken.

FIG. 60 Variation of slip along faults

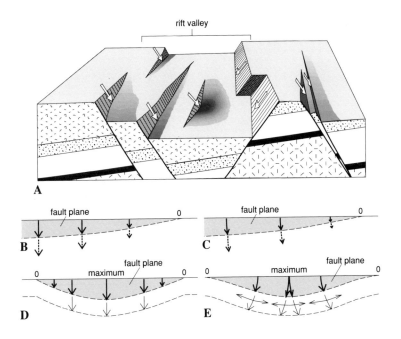

9.6 SEDIMENTATION AND FAULTING

So far we have considered faults that affect rocks formed and consolidated before fracturing took place. However many basins of sedimentation arise as a result of faulting, and in these fault movement can be accompanied by active deposition of sediments. Such situations are often recognised on maps by lateral changes in the types and/or thicknesses of sediments deposited adjacent to and across faults.

In Fig. 61 (diagrams **A** to **C**) development of part of a sedimentary basin is controlled by movement on an extensional fault. Given that sediment may accumulate more rapidly in deeper parts of the basin as movement on the fault proceeds, a situation like that illustrated in diagram **B** could result. Continued fault movement and sedimentation would produce changes in sediment thickness across the controlling fault (diagram **C**). Changes in sediment type, relating to proximity to the fault, could also arise (e.g. **a–b–c** in diagrams **B** and **C**). For example, coarse grained sediments (**b**) could accumulate near the fault, on its downthrow side, and grade laterally away to finer grained sediments (**c**).

If rotation of fault blocks occurs, as illustrated in the sequential diagrams **D** to **F** (Fig. 61), sediments deposited earlier in the history of basin development will be tilted, forming wedges spatially related to the positions of the faults (diagram **E**). Also sedimentary formations will thicken towards the faults (diagram **F**).

Where listric extensional faults control basin development, thickness changes in sedimentary sequences relate not only to rates of subsidence and sediment supply, but also to the curvi-planar shape of the faults. Diagrams **G** to **I** (Fig. 61) illustrate the progressive accumulation of sediment in a basin developed by continuous movement on an extensional listric fault. Abrupt changes in the thicknesses of formations across the fault and gradual increases in their thickness towards the fault, in the hanging-wall, characterise this situation.

Faults which are active during sedimentation are known as growth faults.

FIG. 61 Growth faults

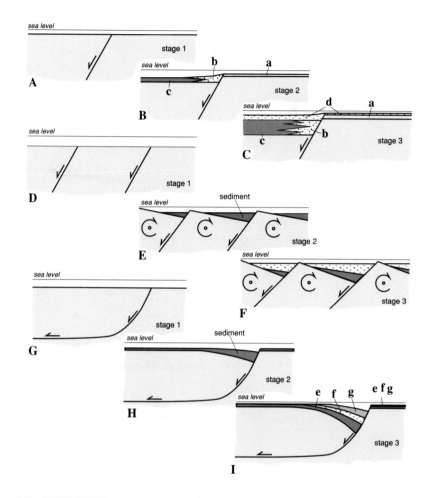

9.7 EXERCISES

9.7.1

Using the relationships between outcrop shape and topography, and structure contours, determine the structure of the area shown in Fig. 62. Construct an accurate cross-section along the line **a** − **b** using the topographic profile provided. Determine the nature of the fault and the relationships between folding and faulting. What is the likely direction and amount of displacement on the fault?

Note that you are given the sequence of sediments and volcanics in order of age, i.e., as a stratigraphic succession (refer back to Chapter 2, section 2.1). Consequently you know the original order in which the formations were laid down, i.e. which are the older, which the younger. In attempting to assess the structure of this map area, as of any other, it is important to locate not only positions of known height for a given surface, directions of dip as indicated by outcrop shape in relation to topography, the positions of the contacts and the fault on the line of section, but also the lines of intersections between the various planar elements, e.g. between the formation boundaries and the fault plane.

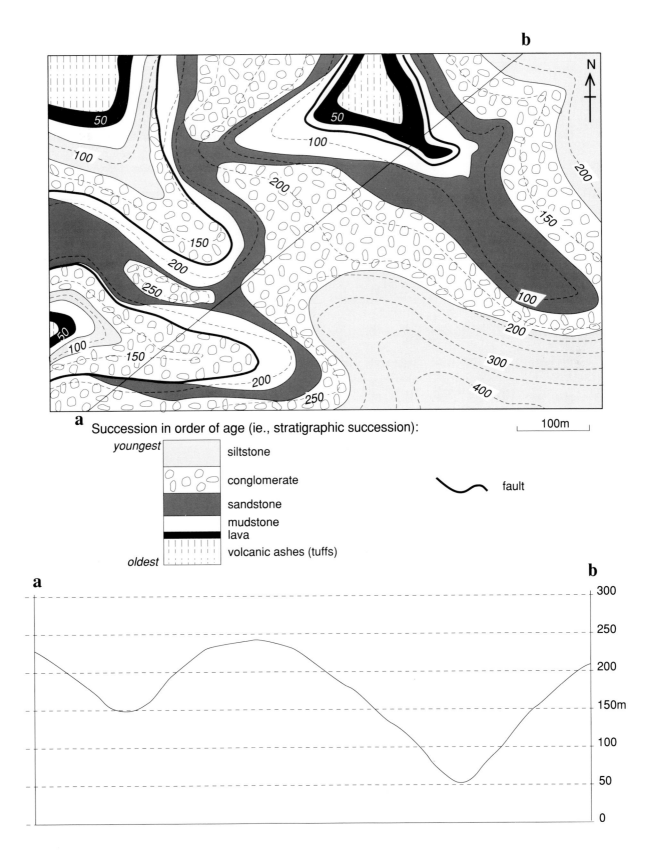

a Succession in order of age (ie., stratigraphic succession):

youngest — siltstone

conglomerate

sandstone

mudstone

lava

oldest — volcanic ashes (tuffs)

fault

100m

FIG. 62 Exercise 9.7.1

9.7.2

By using the relationships between outcrop shape and topography, and structure contours, determine the attitudes of the geological contacts and the faults *throughout the map area* in Fig. 63. Which faults correlate across the river valley? Knowing the stratigraphic succession indicate, on the map, the downthrown side of each fault. On this basis, and before constructing a cross-section, try to determine the types of faults present, and their likely slip directions.

Using all the information on the map construct the cross-section **a–b**. *In doing this remember to locate accurately not only the positions of the contacts, igneous intrusions and faults, but also the intersections of the contacts with the faults, the faults with each other, and the intrusions with faults.* Determine the direction and amount of slip on the faults. What assumptions are you making?

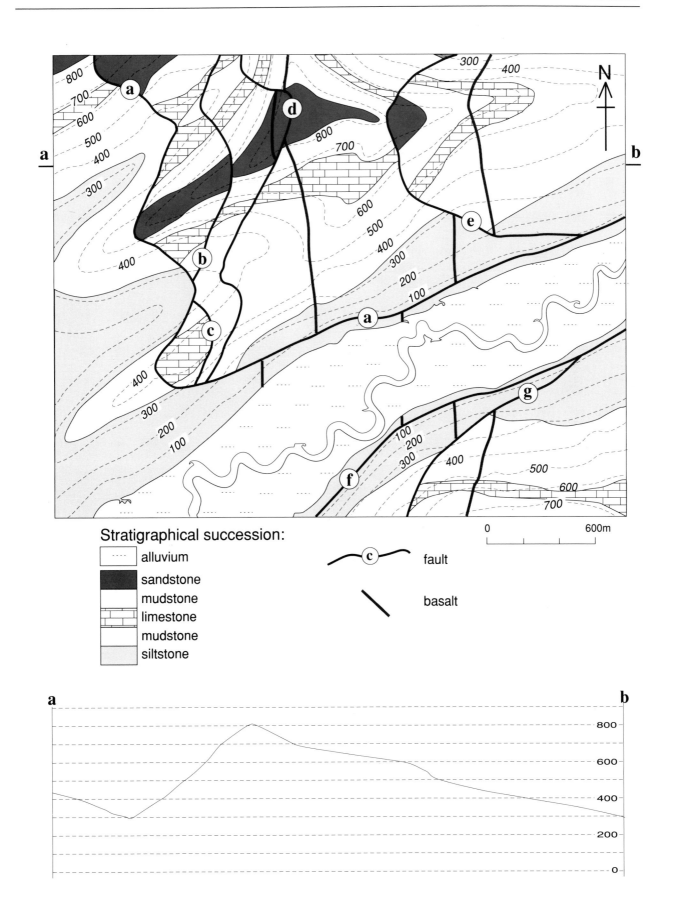

Stratigraphical succession:

- ⋯⋯ alluvium
- ■ sandstone
- mudstone
- limestone
- mudstone
- siltstone

c ⌒ fault

▬ basalt

0   600m

FIG. 63 Exercise 9.7.2

FOLDS

10.1 GEOMETRY OF FOLDS

Whilst most faults reflect the brittle behaviour of rocks under stress and result from crustal compression or extension, most folds record a ductile (plastic) response to crustal compression. They range in scale from microscopic to structures whose dimensions are measured in tens of kilometres. In simple situations we frequently see folds affecting layered sequences of sedimentary rocks which, before folding, lay sub-parallel to the surface of the earth. In the past the terms **anticline** and **syncline** were used to describe the upfolds and downfolds of such sequences (Fig. 64).

Diagram **A** of Fig. 64 illustrates a sequence of sediments, in descending order of age **a** to **e**, which were deposited in a sedimentary basin on a basement of older rocks. Subsequent compression leads to folding and thrusting which cause lateral translation and vertical thickening of the crust (diagram **B**). Note that older sediments are brought nearer to the surface in the cores of upfolds, whereas younger sediments are taken to deeper levels in the cores of downfolds. Where it is known that upfolds are cored by older rocks they are termed **anticlines** and where downfolds are cored by younger rocks they are termed **synclines**. However, because there are situations where downfolds contain older rocks in the core and vice versa, and others where the relative ages of the rocks are not known, the general terms **antiform** and **synform** should be used for upfolds and downfolds respectively.

Diagram **C** gives a closer view of one of the folds in **B** and shows that during the deformation that gave rise to the folds the rocks also suffered internal strain so that grains and minerals became aligned. The planar structure produced by such alignment is called **cleavage**. It is geometrically related to the folds (diagram **C**), a feature that will be discussed later.

The basic nomenclature used in describing folds is summarised in Figs 65 and 66.

A **fold crest** is the line (or area) of maximum topographic height in an antiform; a **fold trough** is the line (or area) of minimum topographic height in a synform (Fig. 65, diagram **A**). **Fold limbs** are shared between adjacent antiforms and synforms (diagram **A**) and can be relatively flat inclined surfaces

FIG. 64 Development of folds

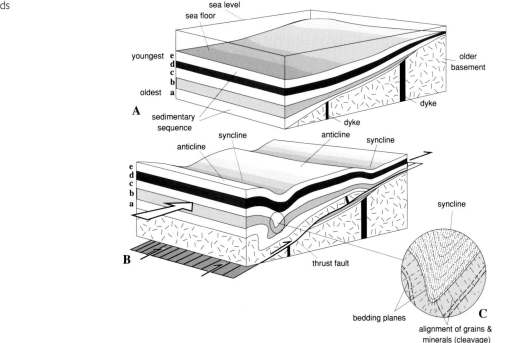

FIG. 65 Fold nomenclature

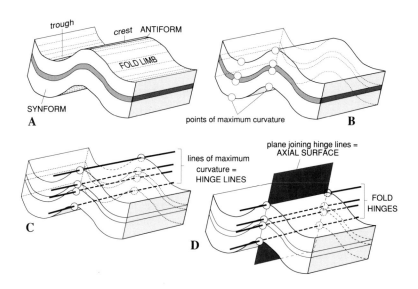

or layers, or they can be sinusoidal. In cross-sections of most folds we can locate **points of maximum curvature** for each folded surface (diagram **B**) and likewise, in three dimensions, lines of maximum curvature, or **hinge lines** (Fig. 65, diagram **C**). A line parallel to the average direction of the fold hinges in a given fold is known as a **fold axis**. It is important to realise in regard to the difference between a fold hinge and a fold axis that in many folds, unlike those illustrated in Fig. 65, fold hinges are not necessarily parallel (see for example Fig. 69). Consequently, whilst hinges may show considerable variation in trend, their average trend (axis) will be a single vector. Fold hinges are features of folds which can be touched and measured in the field whereas fold axes are statistical averages of the trends of several fold hinges.

In referring to the size of folds, the terms *minor* and *major* are often used but it is preferable to measure wavelength and amplitude as in Fig. 66.

Axial surfaces, known also as axial planes or hinge surfaces, are planes which contain all the fold hinges in a given fold (Fig. 65, diagram **D**). They may, as in Fig. 65, be flat inclined planes, and axial surfaces of adjacent folds can be parallel to each other. However, they can also be curvi-planar, they can be non-parallel, and they can occur as conjugate sets (Fig. 67).

The axial surface of a fold is *not* defined as the bisectrix of the interlimb angle, i.e. the angle between the fold limbs. Even though, in special cases, the plane containing the fold hinges will also be this bisectrix, in most cases it is not (Fig. 68, diagrams **A** and **B**). It is also important to note that fold crests and troughs do not necessarily coincide in position with hinge lines, though

FIG. 66 Wavelength and amplitude

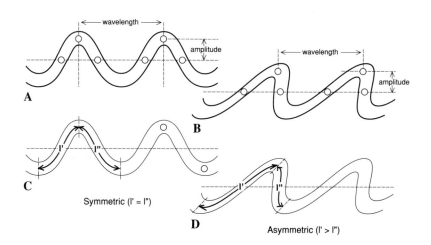

FIG. 67 Fold axial surfaces

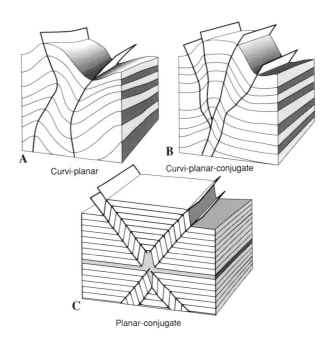

A

Curvi-planar

B

Curvi-planar-conjugate

C

Planar-conjugate

they will have the same trend (Fig. 68, diagrams **C** and **D**).

When fold hinges within a fold are parallel, as in Fig. 69, diagrams **A** and **B**, the fold is referred to as **cylindrical** whereas folds with curvi-linear fold hinges, as in diagrams **C** and **D**, are **non- cylindrical**. Clearly non-cylindrical folds show considerable variation in plunge and shape along their hinge direction whilst cylindrical folds ideally maintain a constant plunge and shape in this direction. Both types of fold can be symmetrical where the lengths of the limbs are equal, or asymmetric where they are not (Fig. 66); they can be upright where the limb dips are equal but opposite, or overturned where the limb dips are in the same general direction. Folds can be open or tight depending on the interlimb angle. Where limbs have equal dip and the same strike, they are referred to as **isoclinal**.

The differences in geometry between cylindrical and non-cylindrical folds are reflected by patterns of structure contours. Thus in Fig. 70 structure contours are regular and parallel in cylindrical, non-plunging folds (diagram **A**) but with non-cylindrical folds (and with plunging, cylindrical folds) structure contours are curvilinear (diagrams **B** and **C**).

FIG. 68 Axial surfaces: crests and hinges

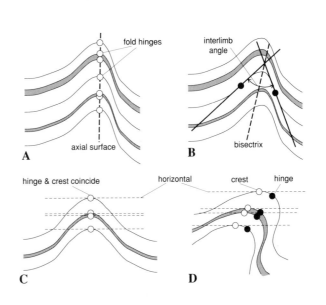

fold hinges

interlimb angle

axial surface

bisectrix

A

B

hinge & crest coincide

horizontal   crest   hinge

C

D

FIG. 69 Cylindrical and non-cylindrical folds

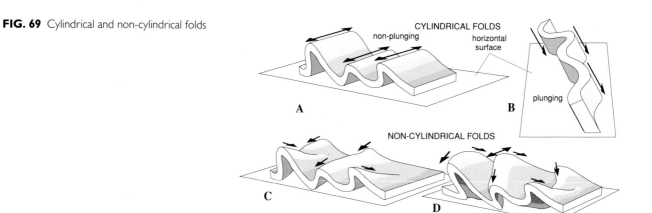

FIG. 70 Folds and structure contours

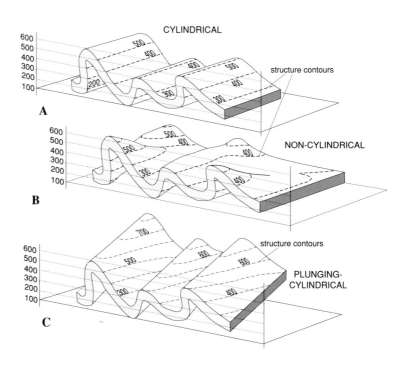

10.2 MAP ANALYSIS OF FOLDS

Analysis of the shapes, trends and geometries of folds from geological maps can be achieved by utilising some of the methods outlined earlier in this manual. For example, in Fig. 71 diagrams **A** and **B** show maps of three geological surfaces folded into a synformal fold with opposed, unequal limb dips of 43° and 67°. Appraisal of outcrop shapes in relation to topography in diagram **A**

FIG. 71 Folds: structure contours and fold hinges

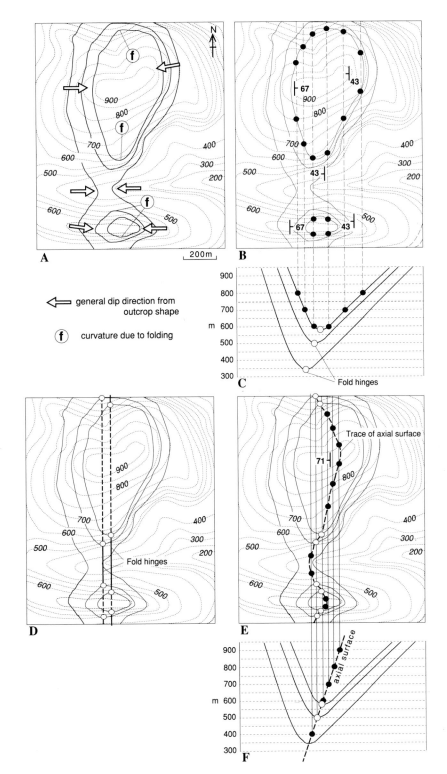

allows us to identify quickly a synformal fold trending more or less North-South with a steeply dipping western, and moderately dipping eastern, limb (arrows in diagram **A**). More detailed analysis of the fold using structure contours, as in diagram **B**, allows us to determine precise dips and, because the structure contours are straight and parallel, all trending N–S, that the fold is cylindrical with a horizontal axis. Construction of structure contours for each of the folded surfaces (not all shown in diagram **B**) also allows an accurate cross-section to be drawn (diagram **C**) and this in turn enables us to locate the positions of fold hinges (note in this case they coincide with the fold troughs).

The positions of fold hinges can also be located on maps, as in Fig. 71, diagram **D**, wherein they are shown as solid lines where they lie above ground surface and as dashed lines where below. Points of emergence of the fold hinges at ground surface are indicated by the open circles. Note that they are located where rapid changes in trend of outcrops occur that are unrelated to topography (**f** in diagram **A**).

The axial surface of the fold can be constructed on the cross-section by joining the points of maximum curvature (i.e. the hinges) for each folded surface; it dips at 71° to the west (Fig. 71, diagram **F**). Because in three dimensions the axial surface contains all the hinge lines which, in this case, are straight and parallel, we can construct its outcrop on the map (diagram **E**). This is known as the **trace of the axial surface** or **axial plane trace**; note that because it is an inclined surface its outcrop 'vees' in the valley in the direction of its dip, as do other geological surfaces.

Analysis of the outcrop patterns of folded rocks can be difficult in areas of complex topography. For example, the map in Fig. 72 shows outcrops of layers of sandstone, mudstone and siltstone of which the attitudes are not readily interpreted. In order to determine the structure we have to decide whether the dips and structure contours are as shown at **x**, **y** or **z**. Note that in cases **x** and **z** the siltstones would *overlie* the mudstones whereas in **y** the siltstones would *underlie* them.

Because in the main valley the outcrops 'vee' (black arrows) in the directions of dip as indicated by structure contours parallel to set **z**, these perhaps appear to be the most likely; but such a proposition requires testing.

Figure 73 shows the patterns of structure contours for the siltstone/sandstone contact produced by using set **x** (diagram **A**) and set **y** (diagram **B**). In constructing these, note that the positions of structure contours are constrained not only by where outcrops cross topographic contours of a particular height but also by the rule:

FIG. 72 Analysis of folds

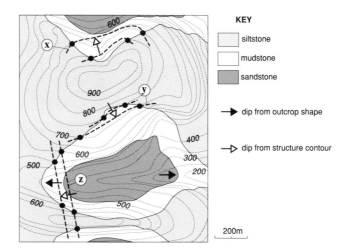

KEY

☐ siltstone

☐ mudstone

▨ sandstone

➡ dip from outcrop shape

⇨ dip from structure contour

200m

FIG. 73 Analysis of folds

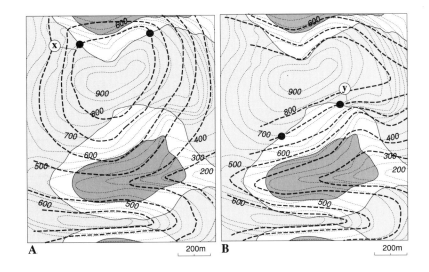

For a given geological surface a structure contour of a given height cannot cross a topographic contour of the same height without that surface cropping out.

Both of the resultant patterns are complex and suggest in **A** that a dome-shaped antiform and an E–W trending, plunging synform coincide in position with the main hill and valley respectively. The sandstones form the core of the synform. In **B** there is a similar coincidence between the position of folds and topographic features; again the sandstones core the synform (try drawing N–S cross-sections for each map).

A much simpler and more regular pattern of structure contours is derived by using set **z** of Fig. 72, as in Fig. 74. This suggests the presence of a single antiformal fold whose hinges are inclined (**plunge**) southwards. Whilst the structure contours are not everywhere straight lines, they form a pattern of regularly spaced similar curves, suggesting that the fold is near-cylindrical but plunging.

Thus whilst the previous explanations are consistent with the data recorded on the map, an interpretation using structure contours of set z is preferred because the fold geometry is least complicated, it does not require the antiform and synform to coincide with topographic features (i.e. the hill and valley respectively), and more importantly it agrees with the relationships

FIG. 74 Analysis of folds

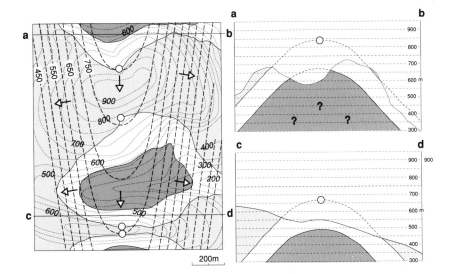

FIG. 75 Fold plunge

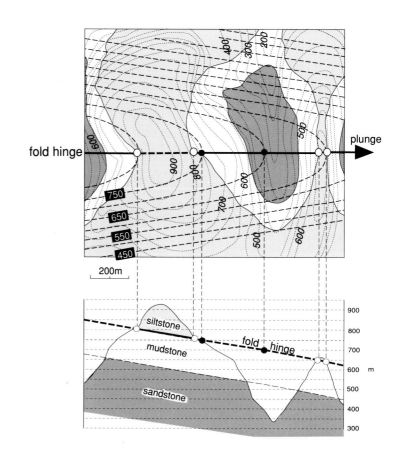

between outcrop shape and topography; it accounts for the 'veeing' of the outcrop in the valley and the curvature of outcrops where there is no simple topographic explanation, i.e. at fold noses. The foregoing discussion also emphasises the importance of assessing relationships between outcrop shape and topography at the outset of any map analysis.

Further investigation of the map, as in Fig. 74, reveals that in the north (section **a–b**) the fold hinge defined by the mudstone/siltstone contact lies at about 840 m whereas in the south (section **c–d**) the same fold hinge lies just above 650 m. From the map we can also see that this hinge intersects ground surface at four places (circles) whose heights are, from North to South, about 825, 775, 670 and 625 m. Thus we can locate several points of known height defining the position of the fold hinge. A cross-section *along* the hinge line illustrates its position and attitude (Fig. 75); the hinge plunges to 180° at 10°.

In recording the trends and attitudes of fold structures on maps we use symbols such as those shown in Fig. 76.

FIG. 76 Fold map symbols

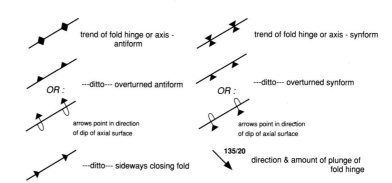

10.3 EXERCISE

10.3.1

The map (Fig. 77) shows the outcrop of the contact between a mudstone and a sandstone formation. Using the relationships between outcrop shape and topography, determine the general attitude of the contact and locate the positions of fold hinge zones. Having thus established the general shape of the contact over the map area, construct structure contours, taking care to base their shape on those for which there exist the largest number of intersections with topographic contours (e.g. that for 300 m). On the basis of the pattern and shapes of the structure contours and points of emergence of fold hinges, draw in the trend of the fold hinges and determine their plunge. Draw a cross-section along the line a–b. Are the folds cylindrical or non-cylindrical?

FIG. 77 Exercise 10.3.1

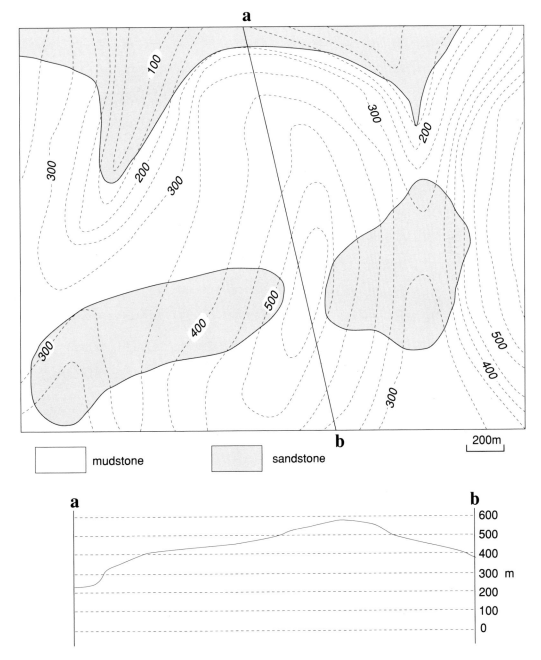

FOLD SHAPE

11.1 ASSESSMENT OF FOLD SHAPE FROM MAPS

The shapes of folds vary considerably depending on such factors as, amongst others: the physical properties of the rock materials during deformation; the amount and type of strain undergone during folding; the orientations of the layering within rocks relative to stress axes (i.e. the directions in which compressive forces are applied).

Because the cross-sectional shape of folds can provide evidence for the deformation mechanisms whereby folds form and affects the way in which we extrapolate fold structures below ground surface, it is important to assess accurately the overall geometry of the surfaces that define folds, in particular to discover whether the folds are sinusoidal or angular and whether variations in layer thickness occur around them. However, such analyses are only diagnostic when undertaken in sections at right angles to the fold hinges (the **profile plane**). Misleading information will be gained in studying sections that are oblique, a point discussed further below.

The folds illustrated in Fig. 78 diagrams **A** and **C** have horizontal hinges; the trends of structure contours and amounts of dip are indicated by the dip and strike symbols. Note that both are overturned. Whilst the dips of the eastern and western limbs in each of the folds are similar, the folds differ in that one is sinusoidal in shape (cross-section **B**), the other angular (diagram **D**): note that the difference in shape is apparent not only in the east–west cross-sections but also from the outcrop shapes on the maps. In section not only are the fold shapes different, but also the thicknesses of the folded layers. When measured at right angles to the layer surface, the orthogonal layer thickness (**t**) is almost constant in **B** whereas in **D**, whilst layer thickness on the fold limbs shows little variation, around the fold nose it rapidly increases to a maximum parallel to the axial plane. Both of these cross-sections, because the folds are cylindrical and have axes that are horizontal and trend north–south, are profile sections; they are therefore also at right angles to the strike of the folded layer surfaces so that the variations in layer thickness are not due to a 'cut effect', i.e. thicknesses are true—not apparent (see Fig. 23).

FIG. 78 Fold shape

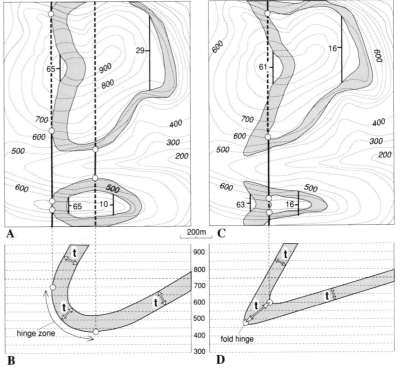

In assessing the geometry of folds, both in the field and from the analysis of maps, it is very important to realise that in sections oblique to the fold axis the 'apparent' fold geometry can be misleading. In Fig. 79 an open, sinusoidal fold wherein layer thickness is almost constant (section **A**) is cut by oblique sections **B** and **C**. In these the limb length appears greater than in the 'true' profile section, the fold appears to be tight rather than open, and layer thickness appears to increase in the fold noses. Thus the fold's 'true' geometry is only revealed in the profile plane **A**.

A rigorous assessment of the geometry of fold structures from data presented on maps, and where available drill-hole logs, etc., is not only of importance in determining possible mechanisms of fold formation but also in predicting the below ground surface location of economic deposits.

FIG. 79 Cross-sections of folds

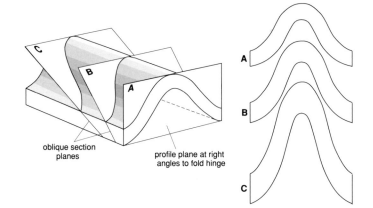

oblique section planes

profile plane at right angles to fold hinge

11.2 EXERCISE

11.2.1

The geological map in Fig. 80 shows the outcrops of a sequence of sedimentary rocks which outside the map area are known to contain an oil-bearing sandstone formation. This is also known to lie below the mudstone marked **a** on the map. Outside the map area the mudstone is on average 250 m thick and oil reservoirs are found in antiformal traps, i.e. in the crestal zones of antiformal folds of the sandstone.

Analyse the structure of the map area, determining the attitudes and locations of fold limbs, axial surfaces, axial plane traces and fold hinges. Using the topographic profile provided, draw an accurate cross-section along the line **a – b**. On the basis of your analysis locate sites for bore-holes which you think likely to strike oil. After you have analysed the map, check your interpretation with that given in Chapter 15.

FIG. 80 Exercise 11.2.1

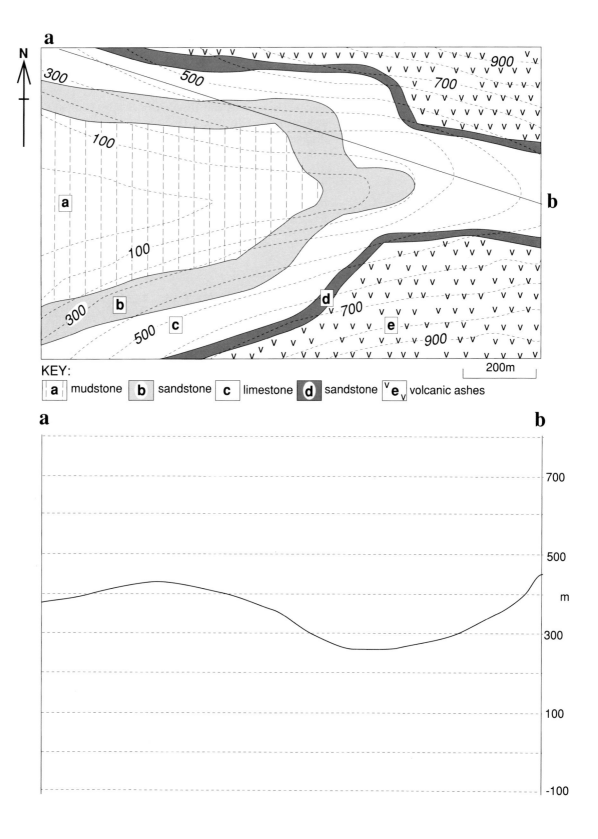

KEY:

| a | mudstone | b | sandstone | c | limestone | d | sandstone | e | volcanic ashes |

11.3 TYPES OF FOLD PROFILE GEOMETRY

Exercise 11.2.1 illustrates the importance of determining the likely profile shapes of folds. In nature these vary considerably but analysis is assisted by considering idealised variations in the profile geometry of curving surfaces. Figure 81 indicates some of the variety we can find in the combinations of shapes of folds seen in sections cut at right angles to their hinges (the profile plane). Diagrams **A** and **B** illustrate angular folds, respectively symmetric and asymmetric, whilst **C** and **D** show examples of sinusoidal folds. The folds in diagrams **E** and **F** are in part angular, in part sinusoidal.

With the exception of diagram **F**, the individual white layers in each illustration were, before folding, of approximately constant thickness. In cases **A**, **B** and **D** this initial constancy of orthogonal thickness (i.e. measured at right angles to the layer surfaces) is maintained after folding, but not in **C** and **E**. This arises because of differences in the ability of the rock materials to 'flow'

FIG. 81 Fold profile geometry

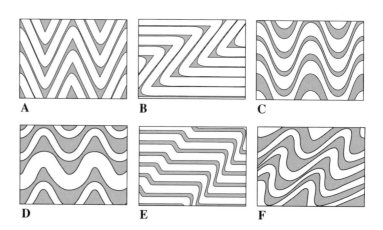

in response to the compressive stresses that gave rise to the folds and differences in the mechanisms whereby the folds formed.

The constant orthogonal layer thickness seen in the white layers of **A**, **B** and **D** defines these folds as having **parallel geometry**—the surfaces on either side of the white layers form parallel curves. In contrast, the curves defining the white layers in diagrams **C** and **E** are the same, and thus they have **similar geometry**. In ideal similar folds constant layer thickness is maintained parallel to the axial surfaces of the folds (Fig. 82, diagram **B**). In the field we find folds, on all scales, that approximate to these geometries but it is important to realise that many folds with intermediate, and combinations of, geometries are common (e.g. Fig. 81, diagram **F**).

Recognition of profile geometry from field data and maps is important because it clearly should affect the way in which we extrapolate folds at depth below ground surface. Similar and near-similar folds can persist for great distances at depth, with the same geometry. In contrast, under certain circumstances, parallel folds are limited. Figure 82, diagram **A**, shows a parallel fold with limb length 1° and constant layer thickness **t**. In the core of the antiform there is excess bed length if we try to keep parallel geometry. Consequently, in many cases with increasing limb dip, a space problem arises and parallel geometry cannot be maintained. There is no such restriction to the

FIG. 82 Parallel and similar folds

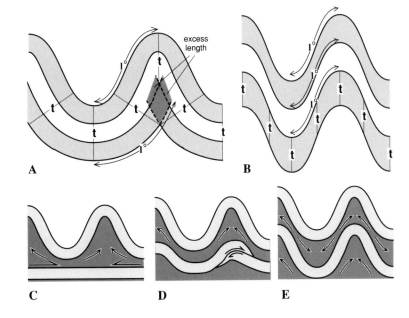

persistence of similar folds at depth (Fig. 82, diagram **B**). With parallel folds the space problem is often overcome by detachment of the folding layers from rocks below (diagram **C**) or by faulting (diagram **D**), combined with flow of the more plastic rock material. Parallel folds can however persist at depth, where interlayers of rock are able to flow to accommodate the developing folds (diagram **E**). In this case notice, the geometry of the surfaces bounding the interlayers is not parallel but intermediate.

In nature we commonly see that fold profile geometry relates to the ease with which rocks of different compositions can undergo plastic flow. When layers are 'stiff' they are termed **competent** and commonly have parallel, or near-parallel, geometry. Layers that can more readily flow (**incompetent**) adopt similar, or intermediate, geometries. Where layers of markedly different physical properties alternate, regularly angular, rather than sinusoidal, folds form. However, it must be realised that the behaviour of rocks undergoing deformation not only depends on rock composition and the way in which layers of different physical properties alternate, but also, on temperature, pressure, and rate of strain, amongst other parameters.

Because of the variations in the profile shapes of folds outlined above, we evidently need, in analysing maps of folded rocks and particularly in constructing cross-sections, to be as accurate as possible in our assessment of profile geometry.

11.4 COMPOSITE CROSS-SECTIONS OF FOLDS

Accurate assessment of fold geometry can only be achieved by rigorous analysis of all the relevant data presented on a geological map. In some instances topographic contours may not be available or may not be accurate so that our analysis of maps cannot rely on structure contours alone. However, topographic data may be given as spot heights and, during geological mapping, height above sea level may be determined using an altimeter. Figure 83, diagram **A**, shows a map of a folded sequence of rock layers cropping out on either side of a flat-bottomed valley with moderately sloping sides; spot heights are given in metres and measurements of the dip and strike of the layering are shown along two traverses, **a–b** and **c–d**.

Inspection of the shapes of the outcrops and the attitude of the layering

FIG. 83 Composite sections: initial analysis

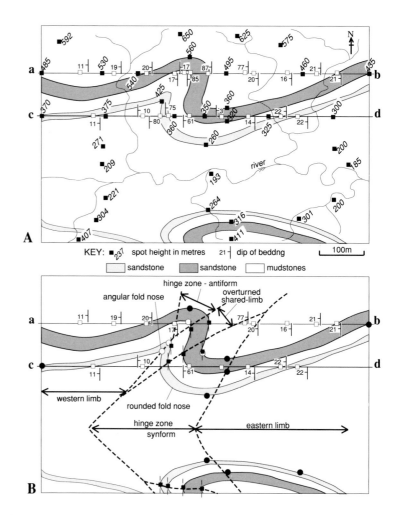

allows us to recognise an asymmetric antiform/synform fold pair with a short, overturned shared limb and long western and eastern limbs (diagram **B**). The shapes of the fold noses and troughs vary from rounded to sub-angular, but overall the outcrop shapes suggest that the folds are sinusoidal rather than angular. The consistent N–S strike, but variable dip, indicates that the fold axes must be horizontal and the folds cylindrical.

Evidently structure contours for the folds cannot be drawn to confirm this because there are no topographic contours. However, diagram **B** shows how we can divide the map into structural sub-areas denoting the approximate boundaries (dashed lines) of fold limbs and hinge zones. It also shows how we can locate, to varying degrees of accuracy, the positions of fold hinges (open circles), and points of known height on the major boundaries (black circles). The black squares with vertical lines show the furthest positions to east and west of the various rock layers on the overturned limb. All of this information is used in constructing a cross-section of the folds.

Figures 84 and 85 illustrate how a cross-section of the folds in Fig. 83 is progressively built up. In Fig. 84, diagrams **A** and **B**, the topographic profiles along both traverses are constructed using the spot heights; preliminary 'form' lines for the geological surfaces are drawn in on the basis of the measurements of dip, both where the boundaries are exposed along the lines of section (grey

FIG. 84 Construction of cross-sections

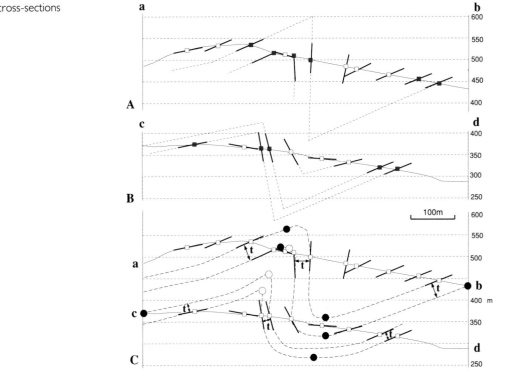

boxes), and in between (white boxes). In **C**, because the folds must be cylindrical (i.e. their profile shape does not vary along the fold axis), the two sections are combined at their appropriate heights and the positions of fold hinges (white circles) and points of known height (black circles) added. Knowing from the map that overall the folds are sinusoidal, the shape of the 'form' lines can be modified. Note also that we can deduce that layer thickness (t), for the two sandstone layers, is approximately constant around the folds. In Fig. 85, the furthest E and W positions for the overturned limb are transferred from the map (black squares), and full account is taken of the outcrop shapes and layer thickness to give a more accurate cross-section. The completed cross-section (Fig. 86, diagram **B**) can now be used to plot the positions of all fold hinges and axial plane traces on the map (diagram **A**).

It is important to note, in this example, how rapid changes in the direction and amount of dip of the layering characterise the hinge zones of the folds. Even where individual rock layers cannot be mapped to demonstrate their presence, these rapid changes allow us to identify readily their positions on the map and section. Note also that the sandstone layers show essentially parallel profile geometry (see Fig. 81).

FIG. 85 Folds – construction of cross-section

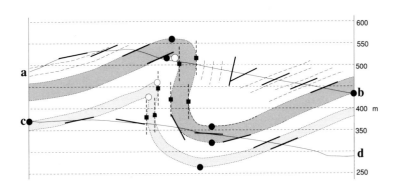

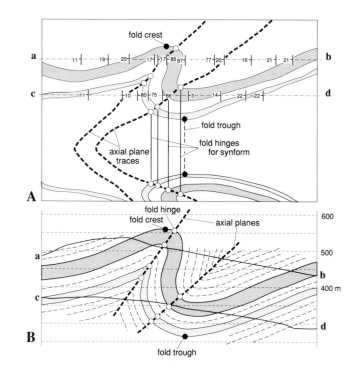

FIG. 86 Folds – location of fold hinges and axial plane trace

11.5 SECTION CONSTRUCTION FOR PLUNGING FOLDS

Where folds are non-cylindrical and therefore their cross-sectional shapes vary along the fold axis, we would be justified in neither projecting data onto a single line of section nor combining different cross-sections. Instead we would have to construct a number of cross-sections spaced along the fold axis, and extrapolate the structure between these sections (Fig. 87).

With plunging cylindrical folds, however, providing we know the angle

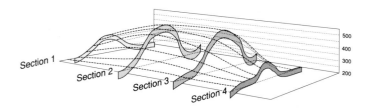

FIG. 87 Sections through a non-cylindrical fold

of plunge, we can still project data from all parts of a map onto a single plane of section. For example, in Fig. 88 (diagram **A**) the outcrop of plunging folds is shown on a horizontal topographic surface.

Points along a folded contact are projected up-plunge onto a plane of section. The positions of each of the points on the plane of section are defined by similar, right-angled triangles, the sizes of which are governed by the angle of plunge (**p**) and the horizontal distance *along the trace of the plunge direction* to the line of section (**d**). We can calculate the height of each point on the section because tan **p** = h/d and therefore **h** = **d** tan **p**. Thus we can determine the position of any point on the map along the line of section and its height above base level.

Where the topography is not flat (as in Fig. 88, diagram **B**) a similar construction can be employed but here the base line height will differ (it will be the topographic height of the particular point chosen for plotting). Thus in diagram **B** the base line for **a** is 275 m, whereas for **e** it is about 210 m.

FIG. 88 Sections of plunging folds

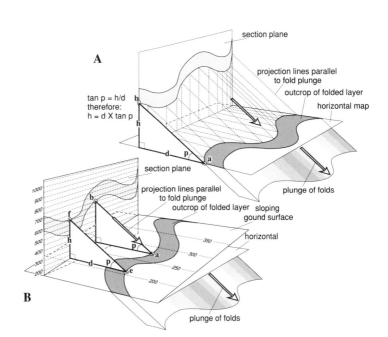

FIG. 89 Profile sections of plunging folds

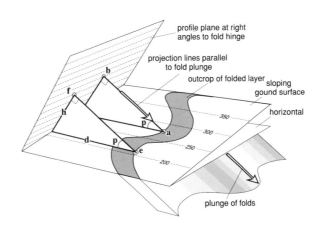

We have already seen that oblique sections through folds give misleading information (refer back to Fig. 79). Where folds are plunging, neither the map nor a vertical cross-section will give the profile shape of the fold. For example, in Fig. 88 the vertical sections are not profile sections: they are not at right angles to the fold plunge. In order to construct a profile section we can modify the construction outlined above, as in Fig. 89. The heights of the selected points **a** and **e**, on the profile plane, are now given by sin **p** = **h/d**; **p** is the angle of plunge, **d** is the horizontal distance from the line of profile in the direction of plunge, **h** is the height on the profile plane.

11.6 EXERCISES

11.6.1

The geological map (Fig. 90) shows the outcrops of four formations and, in a few places, measurements of dip and strike and topographic height. Determine the attitudes of the fold hinges and the geometry of the folds (i.e. whether they are cylindrical or not). Are you justified in projecting the data derived from your analysis and those given as measurements, along strike onto the line **a** – **b**? Taking account of the data and outcrop shapes, as accurately as possible construct a cross-section to illustrate the shape of the folds.

FIG. 90 Exercise 11.6.1

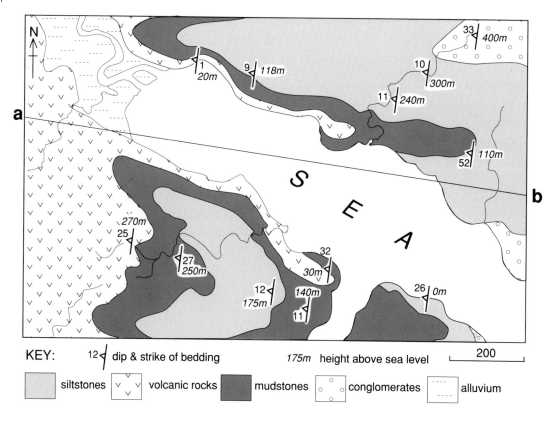

11.6.2

The foregoing exercise emphasises the need to use information from all parts of the map in order to derive a reasonably accurate cross-section, not just data gathered along the line of section. Where the axes of cylindrical folds are horizontal this is straightforward, but, as outlined earlier (Figs. 87 and 88) with non-cylindrical or plunging cylindrical folds, more complex construction techniques have to be utilised.

The geological map in Fig. 91 is of a folded sandstone layer (dark grey) under- and over-lain by mudstones (no ornament). Exposures of rock are outlined and in places show the trends of geological contacts. Readings of the dip and strike of bedding together with the plunge of fold hinges are shown. Given that the area is essentially a horizontal plateau lying at 400 m above sea level, draw in the contacts across the unexposed ground; determine the positions of the various folds, and, using the data from all the exposures of the sandstone, construct accurate vertical and profile sections along the line a–b.

FIG. 91 Exercise 11.6.2

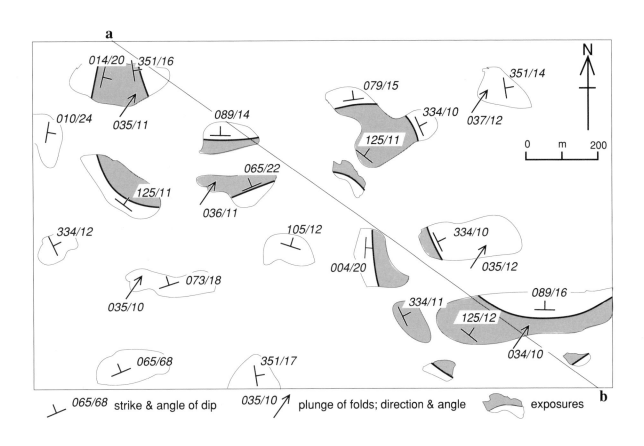

STRUCTURES ASSOCIATED WITH FOLDS

12.1 MINOR FOLDS

During folding, rock layers undergo varying internal strains depending on, amongst other parameters, the physical properties of different rock types i.e. the degree to which they are capable of undergoing plastic flow; the mechanical properties of layered sequences of different materials, e.g. whether or not slip can occur between layers; and layer thickness. Thus at the onset of deformation particularly 'stiff' layers may fold whilst others are able to undergo change of shape by more uniform 'flow' i.e. rather than fold, they thicken at right angles to the direction of maximum compression (Fig. 92, diagrams **A** and **B**). In many cases these differences in behaviour result in the formation of folds of different sizes (**minor** and **major** folds) and, in rocks able to 'flow', the formation of deformation fabrics. The latter, termed **cleavage** and **schistosity**, result from the rotation and/or flattening of the crystals and particles that make up rocks as they deform and, at elevated temperatures, the growth of minerals with a preferred orientation. Such processes impart a grain to rocks along which they are often easily split.

Deformation is progressive in that structures such as folds do not form instantaneously but over long periods of time. Consequently, minor folds formed early in a phase of deformation can become rotated and modified by the formation of larger-scale folds later in the same deformation (Fig. 92). In many cases, because the stress system during a given phase of folding is similar throughout and because the orientation of fold hinges is related to the orientation of stress axes, we find that the geometry and orientation of minor folds mimic that of major. For this reason we refer to such small-scale folds as **parasitic minor folds** (Fig. 92 diagram **C**).

Because we can use observations of the attitudes and geometry of parasitic folds to deduce the positions and nature of major folds which may not be exposed, they are extremely useful both during field mapping and in the interpretation of some geological maps. By noting the profile shapes of minor folds, that is whether they are S, Z or M (or W) -shaped, we can determine which position we are looking at relative to the limbs and noses of major folds (Fig. 93, diagram **A**). Note that M or W-shaped minor folds (i.e. those with equal limb lengths) identify the major fold nose; S-shaped folds (those with unequal limb lengths and anti-clockwise asymmetry) the right-hand limb of the antiform, and Z-shaped (those again with unequal limb length but clockwise asymmetry) the left-hand limb.

If, in the field or on a map, we can identify exposures or areas which, for example, consistently show minor S-folds, we know not only that they lie on

FIG. 92 Parasitic minor folds

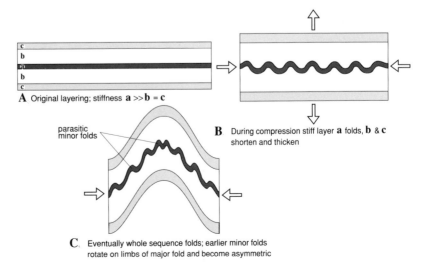

A Original layering; stiffness **a** >> **b** = **c**

B During compression stiff layer **a** folds, **b** & **c** shorten and thicken

parasitic minor folds

C Eventually whole sequence folds; earlier minor folds rotate on limbs of major fold and become asymmetric

the limb of a major fold but also in which direction we would expect to find associated major synforms and antiforms. Thus in Fig. 93, diagram **A**, the S-folds predict not only the occurrence of a major antiform to the left *but also* a synform to the right. It is however very important to realise that fold asymmetry, judged as S or Z, depends on the direction in which we view the minor folds. For example in Fig. 93 (diagram **C**) the same minor fold will have an S shape if viewed along the hinge from the left and a Z shape when viewed from the right. Because of this we must always define the direction in which we view folds. With plunging folds, by convention, this is always in the direction of plunge.

Parasitic minor folds are of further significance because in three dimensions their hinge directions will usually be parallel to those of the major folds to which they relate. Thus by measuring their plunge we can establish the plunge of coeval major folds even though hinges of the latter may not be exposed (Fig. 93, diagram **B**).

12.2   CLEAVAGE AND SCHISTOSITY

During folding many rocks develop cleavage, or at higher temperatures schistosity. When developed at the same time, these planar fabrics and folds are both responses to the same overall stress systems and therefore they are geometrically related. In many cases both fold axes and planes of cleavage (or schistosity) form at right angles to the direction of maximum compression.

Depending on the physical properties of the rocks being deformed, planar fabrics are either parallel to the axial surfaces of folds or are symmetrically disposed about them (Fig. 94, diagrams **A** and **B**). In both cases they are referred to as **axial planar fabrics**. The three-dimensional, geometrical relationships of such fabrics to folds are illustrated in Fig. 94, diagram **C**, which shows that, as well as cleavage planes containing the hinge directions of related minor and major folds, lines of intersection of the folded layer surfaces with the axial planar fabric are parallel to the fold hinges. In exposures in the field such intersections are frequently seen as small-scale linear features on layer surfaces and are known as **intersection lineations**. As with minor fold hinges, they can be used to deduce the orientation of the hinge directions of related major folds (Fig. 94, diagram **C**).

FIG. 93 Geometry of parasitic minor folds

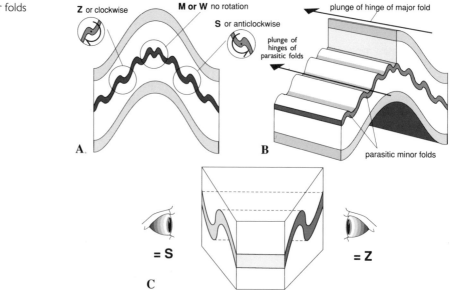

FIG. 94 Axial plane fabrics (cleavage and schistosity)

A. Cleavage parallel to axial surface.

B. Cleavage fans relating to rock type.

axial planar cleavage

plunge of hinge of major fold

axial surface

plunge of hinges of parasitic folds and intersection lineation

intersection of layering and cleavage

C.

cleavage plane

axial plane cleavage

intersection of layering and cleavage

In a similar way to minor folds, the geometrical relationships between axial planar fabrics and folds is of further use in recognising the position of exposures relative to major fold limbs and noses, and thus in predicting the location of larger-scale folds. Figure 95 depicts an exposure of folded sedimentary rocks. This could be on the scale of a roadside exposure or a mountainside. The folds are overturned so that fold limbs dip in the same direction, but note that the dip of the cleavage, relative to bedding, changes from limb to limb of each fold. The overturned limbs are characterised by the dip of cleavage being less steep than bedding, i.e. it lies clockwise of bedding. In contrast, the limbs that are not overturned have cleavage anticlockwise of, and dipping more steeply than, bedding. Fold noses are characterised by the cleavage being at a high angle to the bedding. Thus if in exposures, or in areas on maps, we can determine the attitude of cleavage relative to bedding, we can locate these in relation to larger-scale fold limbs and noses.

12.3 USE OF ASSOCIATED STRUCTURES IN FOLD ANALYSIS

An example of the use of minor folds and cleavage/bedding relationships in locating major folds is given in Fig. 96. Diagram **A** represents observations

FIG. 95 Relationships between cleavage, bedding and folds

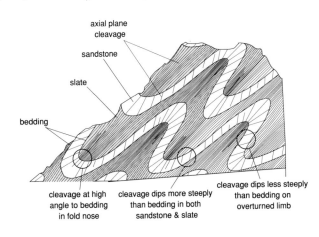

axial plane cleavage

sandstone

slate

bedding

cleavage at high angle to bedding in fold nose

cleavage dips more steeply than bedding in both sandstone & slate

cleavage dips less steeply than bedding on overturned limb

made in cliffs along a coastal traverse through a folded and cleaved sequence of sandstones (white beds) and slates. The field sketches **a** to **e** have been drawn looking northwards at vertical cliff exposures. By accurately recording the profile shapes of minor folds and the relative attitudes of cleavage and bedding we can predict the larger-scale structure shown in diagram **B**. Note in **B** that the attitude of bedding indicated is not that of the minor fold limbs but rather that of a surface drawn through adjacent fold hinges for a particular bedding plane (e.g. diagram **A**, exposure **d**). Such surfaces are known as **enveloping surfaces** and they record the **sheet dip** of the major fold limbs and noses where these comprise trains of parasitic folds.

The foregoing discussion shows that an understanding of the geometric relationships between major and both minor parasitic folds and planar fabrics provides a powerful tool in unravelling the geometry of major folds. This is true both during field mapping and in the analysis of structural maps. Figure 97 is a map of an area of folded rocks showing measurements of bedding and cleavage, and observations of the asymmetry of minor folds for exposures (black circles) along streams.

It is apparent from the map that the bedding, whilst showing some variation, predominantly strikes NNE–SSW and dips towards the WNW. Again with some variation, the cleavage strikes NNE–SSW and dips to the WNW at moderate to steep angles. This suggests that any major folds present will be overturned to the ESE. As the minor folds plunge to the north at around 17° we would expect related major fold hinges to have the same attitude. By noting the relationships between cleavage and bedding and whether the folds are S, Z or W in shape we can locate the limbs and hinge zones of likely major folds. Thus in Fig. 98, diagram **A**, exposures are annotated according to whether they lie on normal (right way up) or overturned limbs, or in hinge zones. Using this information together with that relating to dip and strike of bedding, the approximate boundaries between major fold limbs and hinge zones can be drawn on the map.

The approximate axial plane traces of the major folds can now be located—clearly they will lie in the shaded hinge zone areas (Fig. 98, diagram **B**). In addition, by noting changes in the dip of bedding and the asymmetry of minor folds, we can determine which are anti- and which syn-formal and, provided that the topography is fairly even, we can sketch in form

FIG. 96 Use of minor folds and cleavage/bedding relationships

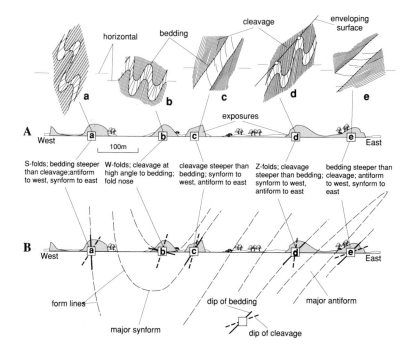

FIG. 97 Fold analysis

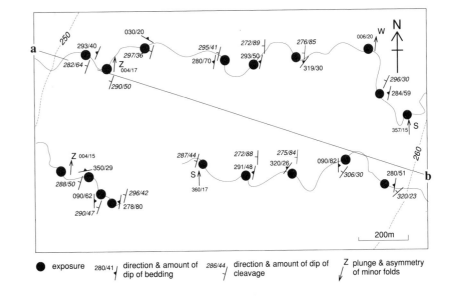

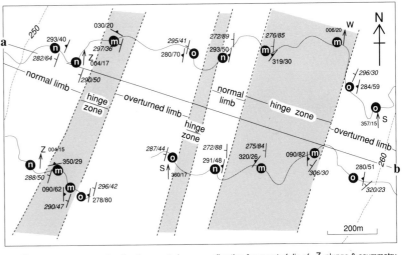

FIG. 98 Location of fold limbs and hinge zones

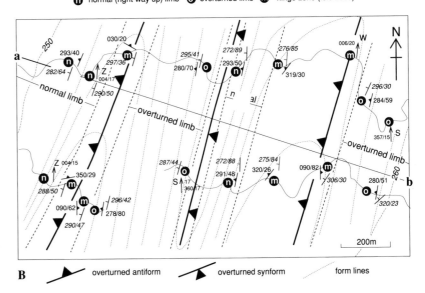

lines showing the approximate traces of the folded layering at the ground surface. Were the land surface irregular we would expect changes in the trend of outcrops relating to topography.

We need to know now whether the folds are cylindrical or not. Because the fold hinges are plunging we cannot readily assess this from parallelism or non-parallelism of the strike of the layering as we have before (e.g. Figs 70 and 83). However, the minor folds all plunge to within a few degrees of due North at about the same angle (between 15 and 20°) and therefore the folds must be cylindrical, at least within the area of the map. Given this, and that the ground surface is even, we can, because the cross-sectional shape of cylindrical folds does not change along the hinge (Figs 69 and 70), project the field data onto the line of section, but we must take account of the angle of plunge.

As with all cylindrical plunging folds, the projected position of any exposure on a section will be a function of the angle of plunge and the distance of the exposure from the line of section *in the direction of plunge* (Fig. 88). Thus in Fig. 99, diagram **A**, the distances of each exposure from the line of section

FIG. 99 Section construction

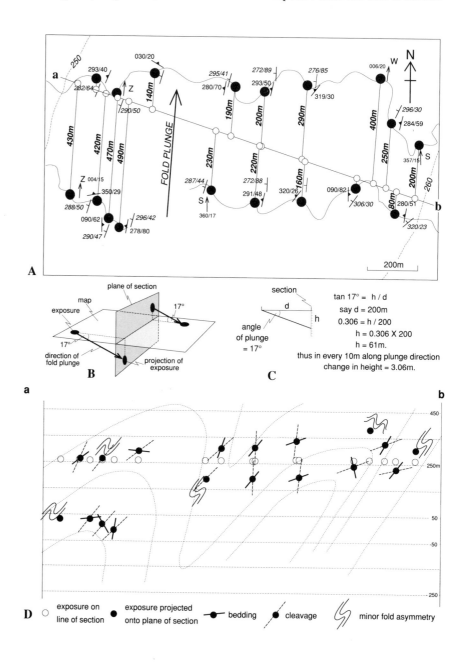

are measured so that their projected heights on the plane of section can be calculated (Fig. 99, diagram C). Note in this example that because the folds plunge northwards, exposures lying north of the section line will project back up the plunge direction to higher levels on the plane of section, whilst those to the south will project down plunge to lower levels (Fig. 99, diagrams B and D). Note in many cases that the line of section is not at right angles to the strike and therefore some angles of dip will be less than those shown on the map; they will be apparent dips.

Using these methods a reasonably accurate cross-section of the folds can be drawn and should incorporate data referring to the attitude of cleavage (note that in this case it fans around the folds), and minor fold asymmetry (Fig. 99, diagram D).

12.4 EXERCISE

12.4.1

The area shown in Fig. 100 comprises folded grey and green slates. Locations of exposures are shown as circles adjacent to which are measurements taken of the attitudes of bedding and/or cleavage. On the basis of the relationships between the dip of bedding and cleavage, determine where the rocks are overturned and where they are not. Wherever possible sketch in the boundaries between areas of relatively low and high dips of bedding, remembering that such boundaries will be close to the axial surfaces of any folds present and therefore will 'vee' over hills and valleys. Also locate the likely positions of hinge zones by noting where the cleavage lies at a high angle to the bedding. With reference to these observations and the known trend of the grey/green slate contact given on the map, draw in the outcrop of this contact and the axial plane traces of all major folds. Again be sure to take into account the effects of topography. Adapting the methods outlined in the analysis of Figures 90 and 96 above, draw in the hinge lines of the folds and indicate which are anti- and which syn-formal. Construct an accurate cross-section along the line a – b showing not only the folded contact but also the positions of axial surfaces and the attitudes of the cleavage.

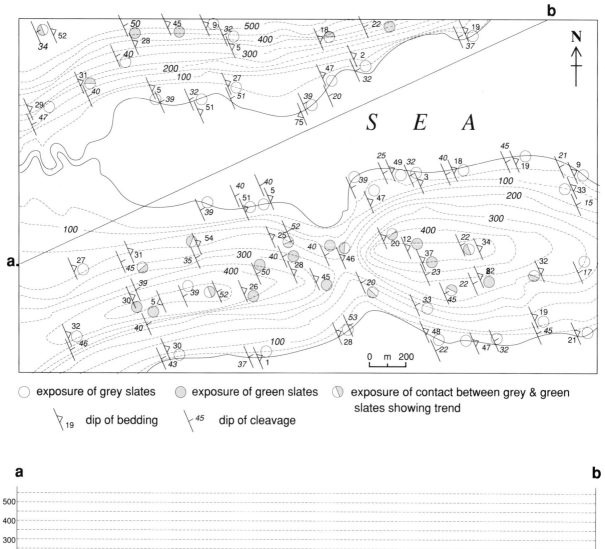

FIG. 100 Exercise 12.4.1

FOLDS AND FAULTS

13.1 DISPLACEMENT OF FOLDS BY FAULTS

The analysis of displacements across faults has been discussed in Chapter 8, wherein it was emphasised that assessment of slip vectors and amounts of slip required recognition of linear elements on either side of fault planes which were, before fault movement, contiguous. In areas of folded rocks, fold hinges provide such linear elements.

In Fig. 101, diagram **A**, a fault **f** cuts and displaces a synformal fold, the position of which is indicated by outcrop shape relative to topography and by dip and strike symbols.

In diagram **B** the attitude of the fault is determined using structure contours; it dips towards 023 at 52°. The approximate positions of the fold hinges can be determined by analysis of the outcrop pattern and its relationship to topography (we look for places where curvature of outcrop or rapid changes in direction are not related to topographic features). To the north the hinge, which must be horizontal because the strike of both fold limbs is the same, lies at about 575 m above sea level, whereas to the south of the fault, whilst still being horizontal, it lies at about 475 m. The block of rocks to the north must therefore have moved up by about 100 m relative to that to the south, or the south down by 100 m relative to the north. The hinge has also been moved horizontally.

By projecting the fold hinge from either side of the fault onto the fault plane we can locate points **a** and **b** which before faulting would have been contiguous (black circles in diagram **B**). From the present positions of **a** and **b** we can therefore deduce that if slip on the fault plane was unidirectional its vector is given by the arrow in diagram **C**. The displacement can be illustrated

FIG. 101 Displacement of fold hinges

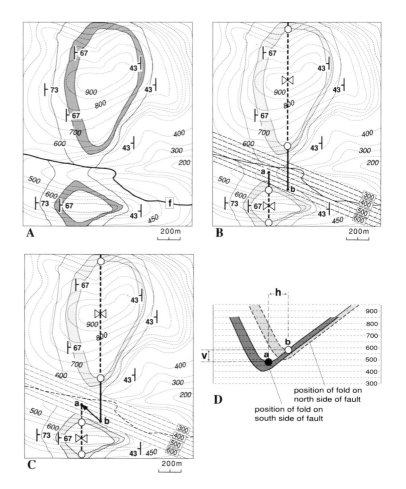

by constructing a vertical cross-section along the strike of the fault plane (diagram **D**). Slip was therefore oblique to both the dip and strike directions of the fault and has caused a vertical separation of 100 m and a horizontal shift of about 130 m (**v** and **h** in diagram **D**). Slip involved either movement of the northern block up to the SE or of the southern block down to the NW. The fault is, therefore neither a reverse nor a strike-slip fault; it has effected oblique-slip. It is very important to realise that the apparently consistent offset of the outcrops of both fold limbs and the hinge line to the west, across the fault plane (diagram **A**), does not prove a component of horizontal slip. As will be discussed later, it could arise by dip-slip. Note, however, that by locating the fold hinge for a given surface (diagram **B**) we can readily deduce a vertical component of slip; the height of the horizontal hinge to the north of the fault is about 575 m but to the south it is about 475 m. There must also be a horizontal component because the hinge is offset across the fault.

The same fold and fault are illustrated in the maps in Fig. 102 but the slip on the fault is different. From the offset of the outcrop pattern across the fault, try firstly to deduce the likely direction and sense of slip.

Has there been a component of horizontal slip or not? Is there any vertical component? As in Fig. 101, locate points on the fault planes that were contiguous before faulting. Calculate both the vertical and horizontal shift on the fault and determine the likely directions and amounts of slip. Check your analysis with that given in Chapter 15.

FIG. 102 Exercise 13.1.1

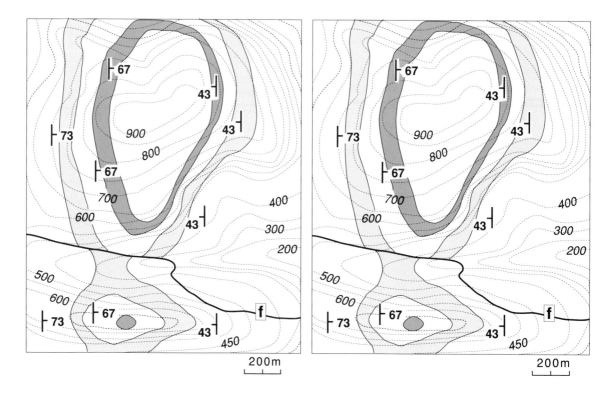

Whilst the location of displaced fold hinges allows us to calculate accurately the slip on faults, folds are also useful in a more general way when trying to assess quickly the affects of faulting. Figure 103, diagrams **A** and **D**, illustrate dip-slip and strike-slip displacement (arrows) of slightly asymmetric antiformal/synformal fold pairs across inclined faults. Diagrams **B** and **E** illustrate the outcrop patterns of these displaced folds after erosion and **C** and **F** are maps of these erosion surfaces. Note that in each case the strike of the faults is at a high angle to the trend of the fold axes. Given certain assumptions, the relationships of the outcrop patterns on either side of the faults, alone, allow us to determine quickly the probable directions of slip on the faults.

In diagram **B** of Fig. 103 note that the down dip displacement across the fault causes, with subsequent erosion, different levels in the folds to be exposed. This is expressed on the map (diagram **C**) by changes in the widths (**w**) of the outcrops of the fold cores across the fault. The synformal core widens to the south of the fault whilst that of the antiform narrows. This situation can only arise by movement of the southern block down relative to the northern (or the northern block up relative to the southern) so that the erosion surface cuts through a higher level in the folds to the south of the fault (see diagram **B**). Because the folds are asymmetric, on the map, the fold axes show a small amount of lateral shift across the fault (a situation that would also arise if the strike of the fault plane were not at right angles to the trend of the fold axes). As explained earlier, such apparent horizontal shifts do not necessarily imply strike-slip or oblique-slip. The absence of horizontal slip in this example is evident because, although the fold limbs and axes show horizontal shift across the fault, this is in different directions along the fault (half-arrows in diagram **C**), a situation that can only arise where the major component of slip is at a high angle to the strike of the fault.

In diagrams **D**, **E** and **F** of Fig. 103 movement on the fault has displaced the folds horizontally along the strike of the fault so that there is no change in vertical level of the folds. This is reflected on the map (diagram **F**) by similarities in the width of outcrops of the cores of the folds, across the fault. Note also that because the erosion surface is flat (topographic relief is low) the amounts of shift of fold limbs and fold axes (open and black half-circles), are constant and in the same sense (half-arrows). This combination of features can be produced only by strike-slip movement.

It is clear from this discussion that, assuming we can identify the same folds on either side of faults and given that topographic relief is low, appraisal of the behaviour of the outcrops of folds across faults can allow us to deduce rapidly the nature of fault displacements without recourse to detailed analysis.

FIG. 103 Displacement of folds across faults

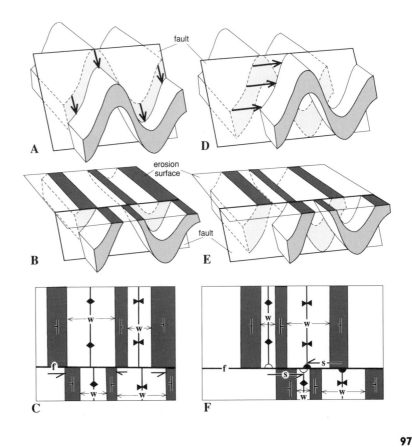

Even where relief is more extreme, similar deductions may be made, but more caution is required because shifts of outcrop may be due to the effects of topography. It is important to remember, however, that because horizontal offsets of outcrops of layers and fold axes across faults can be due to dip-slip movement, or topography, as well as to strike- or oblique-slip, it is the change in width of outcrops, or lack of them, that provides the key evidence in such assessments.

The simple cases discussed above become more complex when folds are overturned. Because in such situations all fold limbs dip in the same direction, albeit at different angles, a superficial appraisal of offsets of fold limbs across faults can lead to erroneous conclusions. In Figure 104, diagram **A**, the strike of the fault is oblique to the axial trend of overturned folds and slip is down the dip of the fault plane. As in Fig. 103, diagram **B**, because the fault downthrows to the south, erosion causes different levels in the structures to be exposed (Fig. 104, diagram **B**). However in contrast to the earlier example, slip causes outcrops of all the fold limbs and the fold axes to consistently shift to the left across the fault (diagram **C**).

At first glance this offset suggests strike-slip, but more careful analysis reveals that the amounts of shift (apparent slip) of individual layers are variable. More importantly, the change in the widths of the cores of the folds across the fault proves a major component of upward or downward movement. The narrowing of the antiformal core to the south, combined with widening of the synform, reveals that there is a vertical component of slip relatively down to the south. However, this evidence alone does not preclude a component of oblique-slip. In order to assess this we would need to find where particular fold hinges intersect the fault plane, as in Fig. 101.

Where the trend of faults is close to the axial directions of the folds they displace, analysis of the amounts and vectors of slip requires construction of the intersections of planar elements (e.g. fold limbs) with fault planes and, as discussed above, calculation of the positions of fold hinges on either side of the faults.

FIG. 104 Displacement of folds across faults

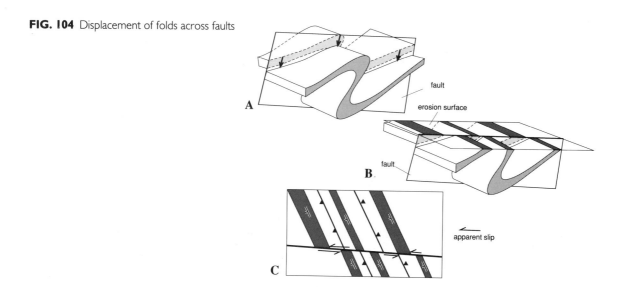

13.2 EXERCISES

13.2.1

The two faults shown in Fig. 105 displace folded sediments for which the vertical sequence is given. Can you tell from the shifts of outcrops across the faults, the directions of slip? By using the relationships between outcrop shape and topography, determine the attitude of the faults and locate, in each fault block, the hinges of the major folds. On this evidence assess the displacements on the faults. Assuming unidirectional movements for each fault, are they dip-, oblique-, or strike-slip faults? Check your deductions by constructing structure contours and comparing the profile geometry of the folds in each fault block.

FIG. 105 Exercise 13.2.1

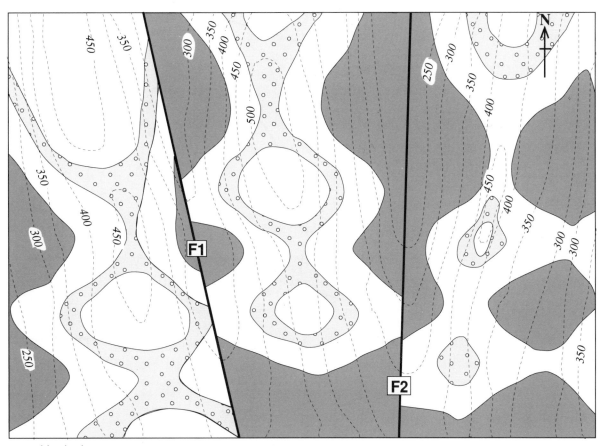

Vertical sequence:

mudstone

sandstone

mudstone

siltstone

200m

13.2.2

The area illustrated in Fig. 106 is one of relatively low topographic relief except for deeply incised river valleys. Repeated outcrops of the same formations of siltstone, sandstone and conglomerate are displaced by faults with differing attitudes. On the basis of the symbols for dip and strike, outcrop shape, and the repetition of outcrops of the sedimentary formations, locate and determine the types of major folds present. Using the effects of faulting on the folds, assess, as far as possible, the nature of the faults and indicate probable directions and amounts of slip. Explain why the displacement of the folds is variable across faults 1 and 4. (*You may find it helpful to refer back to Chapter 8 and earlier parts of Chapter 13 before starting your analysis.*)

When you have completed your analysis check your conclusions with the solution and explanation given in Chapter 15.

FIG. 106 Exercise 13.2.2

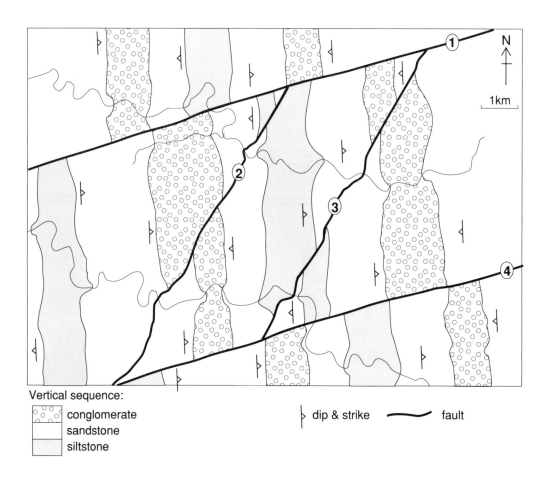

Vertical sequence:

conglomerate
sandstone
siltstone

dip & strike   fault

14

MAP EXERCISES

The following sequence of map exercises can be completed by using a combination of the techniques that have been explained in the earlier parts of this manual. In some cases more advanced techniques of analysis (such as the use of stereographic projections of data) could be applied but they are beyond the scope of this introductory manual. Whilst the maps are not of actual areas in Britain or elsewhere, many are based on real situations. Thus they provide exercises which will enable the reader to gain experience that is applicable in the analysis of geological maps and to develop further the ability to derive a three-dimensional appreciation of geological structures as expressed on maps. As with previous exercises, solutions are provided in Chapter 15 which, whilst only being given in summary, will help you to check your analysis and understand how to apply the techniques outlined in this manual. They will also show you how the geological history of a given area can be unravelled. In each case you should fully, and as accurately as possible, analyse the structures present and deduce a geological history.

FIG. 107 Exercise 14.0.1

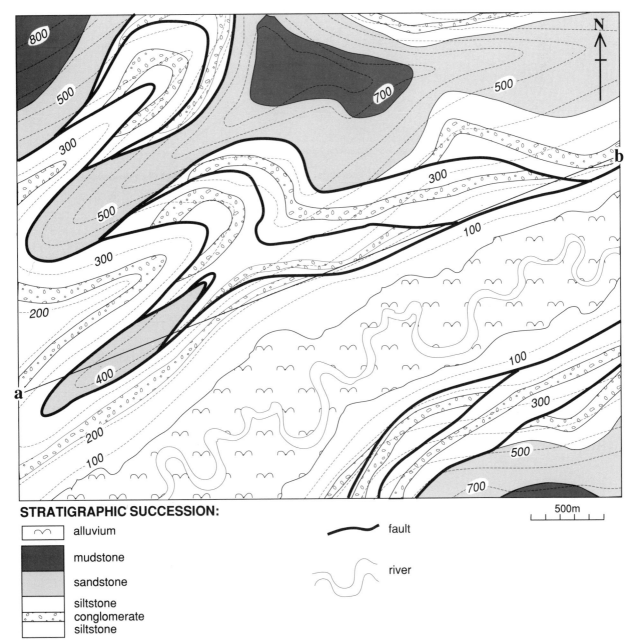

STRATIGRAPHIC SUCCESSION:

alluvium	
mudstone	
sandstone	
siltstone conglomerate siltstone	

fault

river

500m

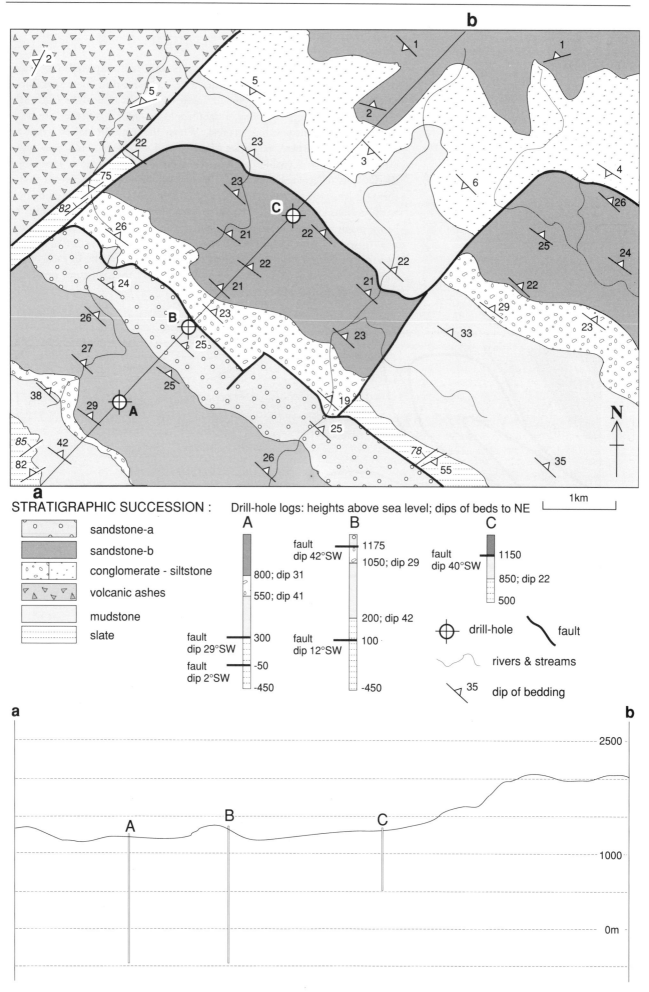

STRATIGRAPHIC SUCCESSION :

Drill-hole logs: heights above sea level; dips of beds to NE

- sandstone-a
- sandstone-b
- conglomerate - siltstone
- volcanic ashes
- mudstone
- slate

A
fault dip 29°SW — 300
fault dip 2°SW — -50
-450
800; dip 31
550; dip 41

B
fault dip 42°SW — 1175
1050; dip 29
200; dip 42
fault dip 12°SW — 100
-450

C
fault dip 40°SW — 1150
850; dip 22
500

- ⊕ drill-hole
- fault
- rivers & streams
- dip of bedding 35

1km

FIG. 108 Exercise 14.0.2

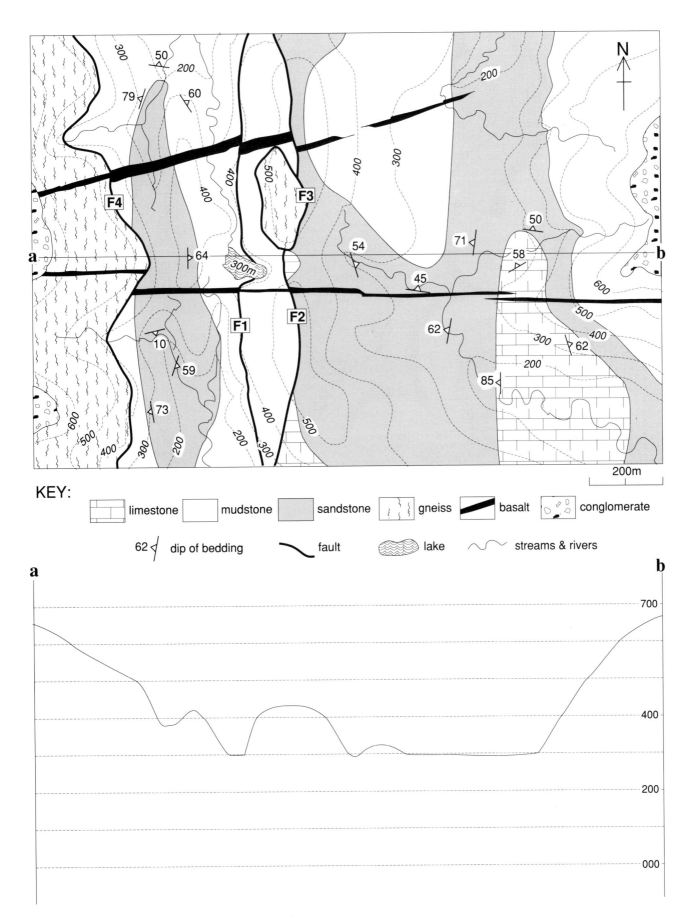

FIG. 109 Exercise 14.0.3

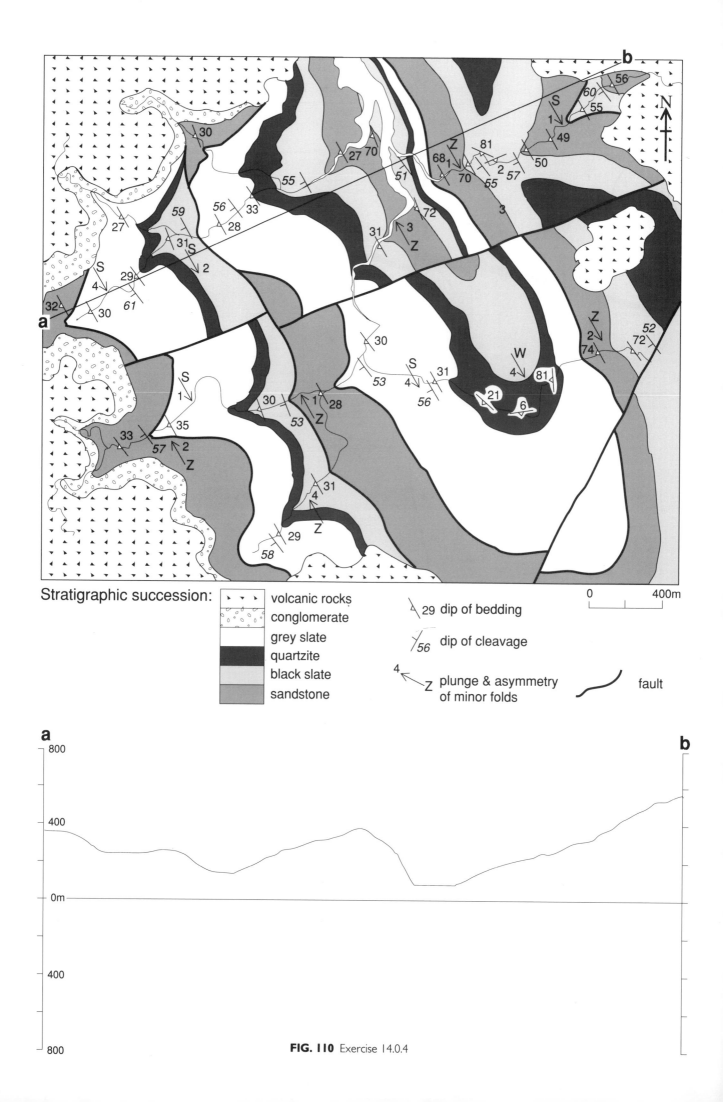

Stratigraphic succession:

volcanic rocks
conglomerate
grey slate
quartzite
black slate
sandstone

29 dip of bedding

56 dip of cleavage

4 Z plunge & asymmetry of minor folds

fault

0   400m

FIG. 110 Exercise 14.0.4

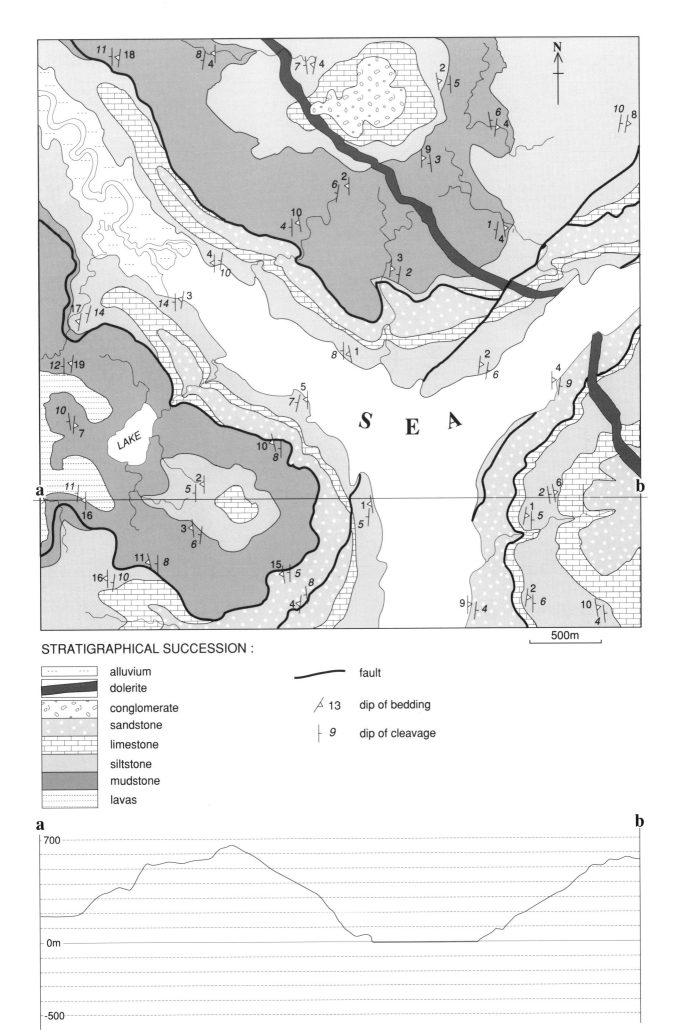

STRATIGRAPHICAL SUCCESSION :

	alluvium
	dolerite
	conglomerate
	sandstone
	limestone
	siltstone
	mudstone
	lavas

——— fault

⋏ 13   dip of bedding

⊢ 9   dip of cleavage

500m

FIG. 111 Exercise 14.0.5

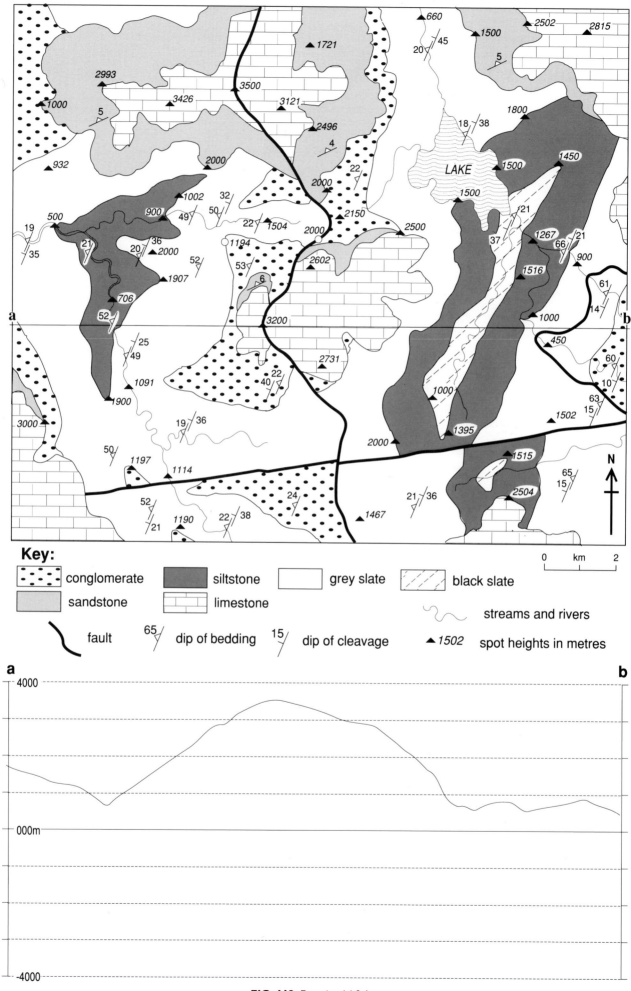

FIG. 112 Exercise 14.0.6

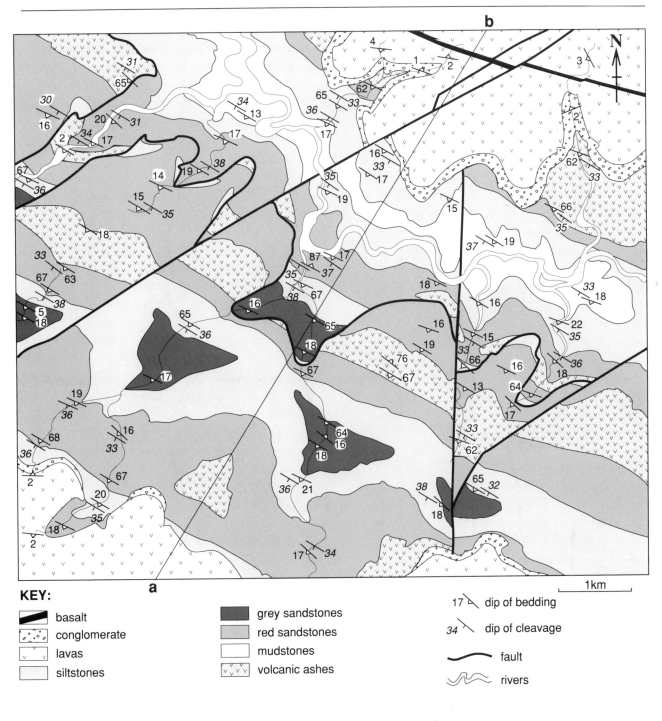

KEY:

▰ basalt		▰ grey sandstones	
⁘ conglomerate		▰ red sandstones	
∨ lavas		☐ mudstones	
☐ siltstones		⩒ volcanic ashes	

17 ⟍ dip of bedding

34 ⟍ dip of cleavage

⌇ fault

⌇ rivers

1km

a ——————————————— b

1500
1000
500
0 m
500

FIG. 113 Exercise 14.0.7

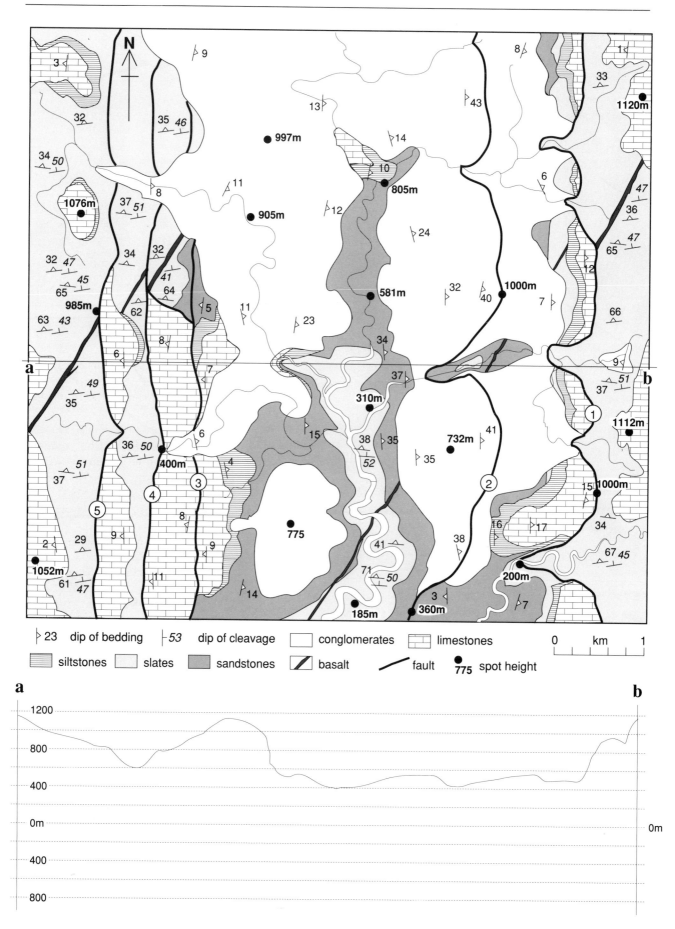

⊳ 23 dip of bedding   ⊢ 53 dip of cleavage   ☐ conglomerates   ▦ limestones   0   km   1

▤ siltstones   ☐ slates   ▨ sandstones   ◩ basalt   ━ fault   ●775 spot height

FIG. 114 Exercise 14.0.8

15

SOLUTIONS TO EXERCISES

Exercise 3.7.1A

The lines between the different rock types on the map represent the surface outcrops of the contacts between them. Points of known height on the outcrops of the sandstone/mudstone and siltstone/sandstone contacts are indicated in diagram **A** below, by white circles. However, no such points are present for the conglomerate/sandstone contact, though we can see that this lies everywhere at between 400 and 500 m. Structure contours drawn through points of like height for the siltstone/sandstone and the sandstone/mudstone contacts are straight and parallel but those for the latter are more narrowly spaced than the former. Thus these contacts are flat, though inclined, planes which strike in the same direction but have different angles of dip to the east (taking north as being towards the top of the page).

In order to draw a cross-section, positions where the structure contours cut the line of section are marked on a slip of paper laid along the line of section (diagram **B**) and are transferred to the cross-section (diagram **C**). Points where outcrops of the contacts cross the line of section (square boxes) are likewise located on the cross-section. In this way the positions of the two contacts both above and below ground level can be established.

It is apparent on the map that the conglomerate/sandstone contact joins that of the siltstone/sandstone at the points marked by black circles. These lie on a straight line which intersects the line of section and therefore the point where the two surfaces join can be transferred onto the cross-section (black circle in diagram **C**). Connecting this point to the point on the section where the conglomerate/sandstone contact crops out, shows that this contact is horizontal. Such an attitude is also indicated by the parallelism of its outcrop with the topographic contours; this because topographic contours are, by definition, the intersections of horizontal planes of a given height with the land surface. From the section we can deduce that the vertical sequence of rock formations is as shown, with the base of the siltstone truncating the rocks beneath. Such a junction is said to be disconformable. However, from neither the map nor the

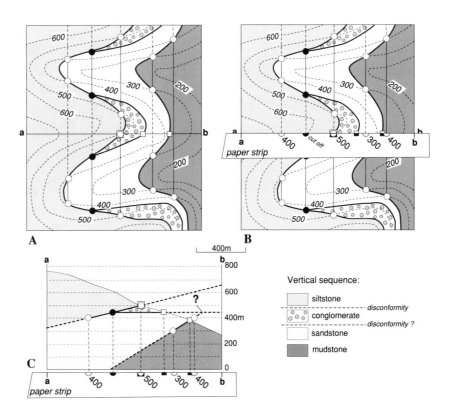

section can we deduce whether or not the conglomerate/sandstone contact is disconformable with that of the sand/mudstone, or vice versa.

Exercise 3.7.1B

Some of the exposures of the coal seam (black circles on the map) lie at topographic contours—three of them at 200 m (diagram **A**). The line joining these is straight, suggesting that the seam strikes NW–SE (taking north as being towards the top of the map). Similarly a line through the two exposures lying at 300 m trends NW–SE parallel to that at 200 m. This indicates a consistent strike for the coal seam. A parallel line drawn through the 400 m exposure is at the same distance (**s**) from the 300 m structure contour, indicating a uniform dip towards the SW. The 600 m exposure lies at twice this distance from the 400 m extrapolated structure contour. Thus the attitude of the coal seam is consistent with it being a flat but inclined plane dipping at a constant angle to the SW. Because we now know the direction of strike and the angle of dip we can extrapolate the positions of the 500 and 100 m structure contours (they will be at distance **s** from adjacent structure contours).

Because by definition any geological plane must be very near the topographic surface where structure and topographic contours of the same height meet, we can plot on the map extra 'exposures' (open circles), i.e. places where the outcrop is covered by soil or vegetation. By 'joining the dots' we can now construct a map as though there were no soil cover (diagram **B**). However, in doing this we have to take account of the effect of topography on outcrop shape, i.e. 'veeing' across valleys and ridges. As the rocks dip to the SW, the rocks overlying the coal seam occupy the high ground as shown in diagram **B** and the cross-section (diagram **C**).

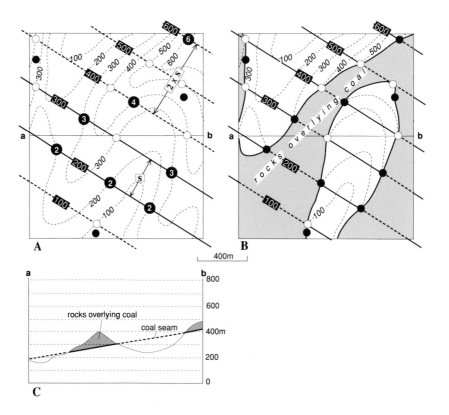

Exercise 3.7.2A

The limestone occupies the high ground and its contact with the mudstone is parallel to the topographic contours. The limestone therefore overlies the mudstone and its contact is approximately horizontal, lying between 600 and 700 m. The shape and attitude of the mudstone/sandstone contact is difficult to interpret and it appears that two sets of structure contours can be drawn: a complex set trending approximately E–W but in some places N–S, OR a consistent parallel set, trending N–S (respectively, dashed lines and solid lines in diagram **A**). The latter when completed (diagram **B**) offers a simple explanation of the shape of the contact and accords with the 'veeing' of the outcrop over the ridge. The contact therefore appears to be curvi-planar with the mudstone overlying the sandstone and the contact in the east dipping towards the west, and in the west dipping towards the east; the change of attitude taking place across a N–S line through the centre of the map (circle in cross-section). Thus the contact is folded and is cut across disconformably by the horizontal limestone (diagram **C**).

The alternative set of structure contours suggests an extremely complex shape for the contact with highly variable dip and strike. Whilst such a shape is not impossible, the existence of the much simpler explanation outlined above, and the relationships between outcrop shape and topography, make it very improbable.

From the section we can see that the vertical sequence is as shown and that the base of the limestone formation is disconformable with the underlying rocks, i.e. it has a different attitude and it must have formed after the older rocks were folded.

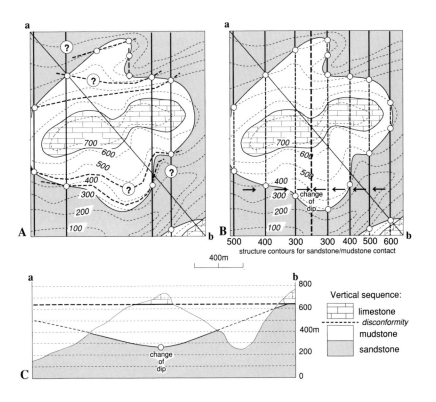

Exercise 3.7.2B

The mudstone forms the ridges and thus it lies above the sandstones. The mudstone/sandstone contact consistently drops in height southwards, suggesting a dip in this direction.

With the exception of three points at 400 m and at 600 m it is not possible to draw straight-line structure contours through more than two points of the same height (diagram **A**). However, the provisional structure contours in diagram **A** have the same overall trend and fit with the southerly drop in height of the contact. Curved structure contours, as in diagram **B**, are however consistent with the data and show that the contact is curvi-planar with variable dip and strike.

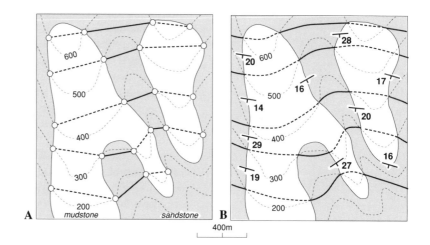

Exercise 4.4.1

Outcrop shapes across the valley suggest dips, for both surfaces, in a westerly direction. However, in the north, outcrops of the contacts cross topographic features with little change in direction, suggesting steep dips to the south. For surface **s** (diagram **A**) there are three points of like height at 350 and 400 m respectively (black circles). These do not lie on straight parallel lines but rather on similarly shaped curves. The surface must therefore be curvi-planar with variable dip and strike. These two structure contours are used, together with the other points of known height (white circles), to extrapolate the structure

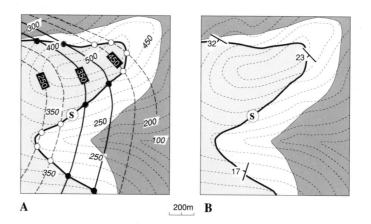

contours for 250, 300, 450 and 500 m. Values for dip (diagram **B**) can be calculated from the spacing of the structure contours using the technique outlined in Fig. 17.

For surface **r** (diagram **C**) there are five points lying at 350 m and again these define a curved structure contour. Other data points are limited, but those at 200, 250 and 300 m define contours that trend in the same direction as that for 350 m. Likewise the two points at 400 m fit a similar curve. The surface is again curvi-planar with dips as shown in diagram **D**.

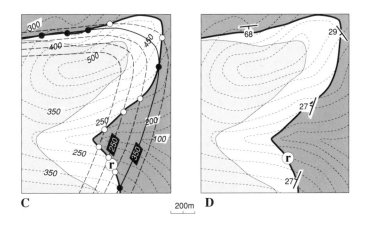

C

200m

D

Exercise 4.4.2

The outcrop shapes in diagram **A**, in relation to topography, suggest dips generally to east and west (arrows). The curving outcrops on the valley sides at **f** cannot be explained by topography because the valley sides are reasonably flat, though inclined, surfaces. Both of these observations show that the geological surface is curvi-planar such that a downfold (synform) trends N–S across the map. Construction of structure contours and the cross-section confirm this. Note that the two sides of the fold dip at different angles. The fold is therefore asymmetric in shape.

In diagram **B** similar observations reveal a comparable structure to that in **A** but the dips on the western side of the fold are vertical (the outcrops trend in a straight line across the topography; refer back to Fig. 22, diagram **E**).

The highly sinuous outcrops in diagram **C** suggest overall low to moderate dips to the west (see Fig. 22, diagrams **B** and **C**). However, in the floor of the valley to the west, outcrop curvature indicates a dip to the east (arrow). The acutely curving outcrops of the siltstone/mudstone contact on the sides of the valley at **f** must relate to the presence of a fold (for the same reasons as cited for **A** above) but here both sides of the fold dip in the same direction. Construction of structure contours and calculation of dips show that the western side of the fold dips at 45° whilst the eastern dips at 10° (see cross-section). The fold is therefore asymmetric and overturned.

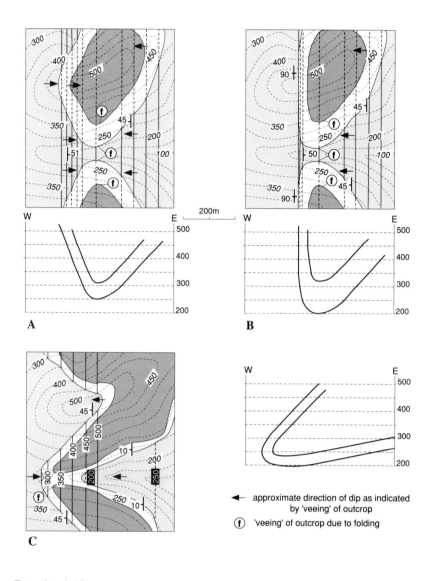

A

B

C

← approximate direction of dip as indicated by 'veeing' of outcrop

ⓕ 'veeing' of outcrop due to folding

Exercise 4.4.3

South-westwards from the NE corner of map **A**, the structure contours gain height to 350 m and then progressively drop to 200 m. Thus they define an upfold (antiform) trending from NW to SE. Given that a geological plane will be at ground level where structure and topographic contours of the same value cross, the circles (diagram **A**) locate the trace of the plane on the ground surface. Taking account of topography the outcrop can be drawn in and the 'vees' in the valleys and over ridges (arrows) relate to the south-westerly and north-easterly dips of the two sides of the fold whilst the curving outcrops at **f** locate the fold closure.

In diagram **B**, from NE to SW, the structure contours progressively lose height to 350 m, then gain height to 450 m, lose height to 300 m and finally gain height again to 350 m. The surface is folded into two synforms with an intervening antiform.

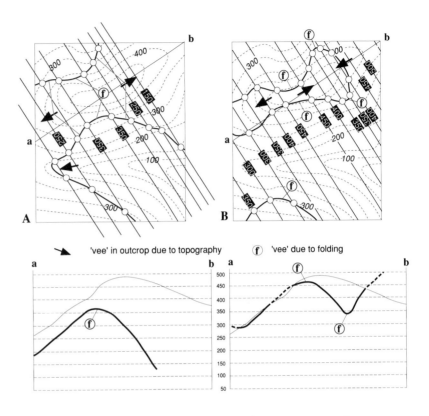

Exercise 4.4.4

Because the area shown in the map is large and the topographic relief is moderate, minor fluctuations in outcrop shape due to topography are not seen. However, major affects, e.g. in relation to river valleys, hills and ridges, are apparent and these can be utilised to determine the general attitudes of the geological surfaces present (see annotations on map **A**). Whilst the cross-section cannot be accurate (structure contours cannot be constructed) it nevertheless gives a useful appraisal of the structure of the area and the timing of geological events. A lower sequence of silts, sands and lavas was laid down and then folded. Some time later dykes were intruded (they cut across the folds) and both these and the older sequence were then displaced across the fault (outcrops of the dykes and the folds are offset).

The base of the conglomerate is disconformable and, as it is not marked as a fault, it must be of sedimentary origin. This type of contact arises because erosion has planed off both the older rocks and structures and then the conglomerates were deposited on this erosion surface. Disconformities of this kind are known as **unconformities**.

Note in the section **B** that the geology is extrapolated both above and below ground surface and the truncation at the unconformity is shown. The vertical sequence of layered formations is, from lowest to highest, siltstone–sandstone–lava–conglomerate, with the dolerite intruding the first three.

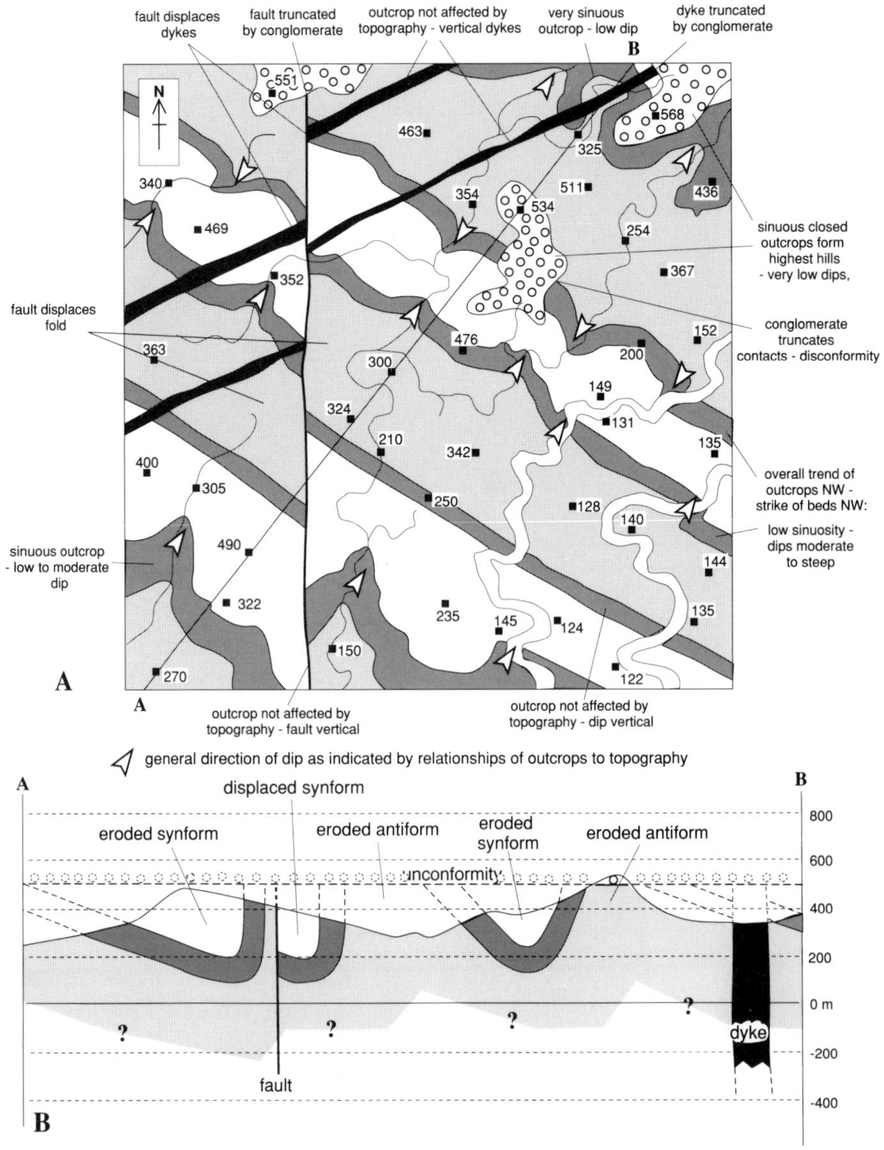

fault displaces dykes
fault truncated by conglomerate
outcrop not affected by topography - vertical dykes
very sinuous outcrop - low dip
dyke truncated by conglomerate
B
N
sinuous closed outcrops form highest hills - very low dips,
fault displaces fold
conglomerate truncates contacts - disconformity
overall trend of outcrops NW - strike of beds NW:
low sinuosity - dips moderate to steep
sinuous outcrop - low to moderate dip
A
A
outcrop not affected by topography - fault vertical
outcrop not affected by topography - dip vertical
general direction of dip as indicated by relationships of outcrops to topography
A
B
displaced synform
eroded synform
eroded antiform
eroded synform
eroded antiform
unconformity
800
600
400
200
0 m
-200
-400
?
dyke
fault
B

Exercise 5.3.1

The limestone occupies the tops of the hills and therefore overlies the mudstones and sandstones, truncating the contact between them. Structure contours show that the limestone dips to the west at a low angle (dashed lines) whereas the contact between the mudstones and sandstones (solid lines) dips to the south-south-west at a moderate angle. To locate the intersection between two geological surfaces we need firstly to examine the map for direct evidence. There are four points where the mudstone/sandstone contact meets the base of the limestone (white circles). These lie on a straight line which marks the position of the intersection. The intersection can also be located by noting where structure contours of the same height for both surfaces cross (black circles). All of these points lie on a straight line because, as can be deduced from the structure contours, the two surfaces are flat, though inclined, planes (see Fig. 33).

The attitude of the line of intersection is given by the trend of its trace on the map (072 to 252) and by the spacing of points of known height along it, i.e. the black circles. These fall in height from ENE to WSW, from 600 to 400 m and therefore the angle of plunge (24°) can be calculated as in Fig. 32.

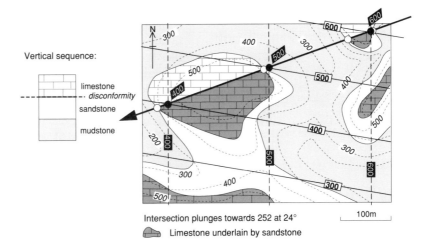

Vertical sequence:

limestone
disconformity
sandstone
mudstone

Intersection plunges towards 252 at 24°

Limestone underlain by sandstone

100m

Exercise 6.3.1

Along straight lines drawn on the map connecting bore-holes **A**, **D** and **C**, the

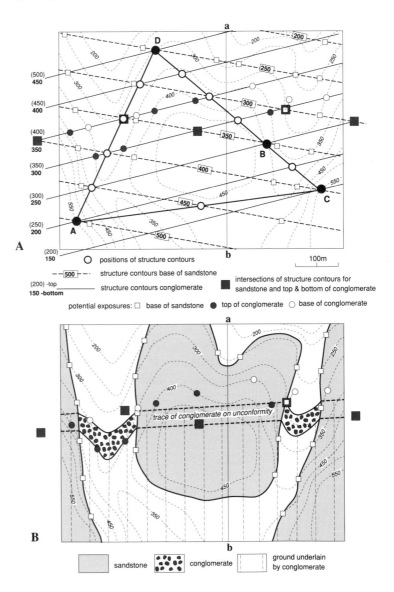

○ positions of structure contours

--- 500 --- structure contours base of sandstone

(200) -top
150 -bottom
structure contours conglomerate

■ intersections of structure contours for sandstone and top & bottom of conglomerate

potential exposures: □ base of sandstone ● top of conglomerate ○ base of conglomerate

trace of conglomerate on unconformity

sandstone · conglomerate · ground underlain by conglomerate

base of the sandstone drops, from south to north, from 500 to 250 m (**A** to **D**); from 400 to 250 m (**C** to **D**). Between **A** and **C** it drops to the ENE from 500 to 400 m and between **C** and **B** there is a 50 m difference in height. Dividing the distances between these points into appropriate divisions to give 50 m intervals, allows the structure contours for the base of the sandstone to be constructed (dashed lines in diagram **A**). Note along line **D** to **C** that the intervals are the same as distance **B** to **C**, which we know from the bore-hole log relates to a 50 m difference in height. These constructions show that the surface can be interpreted as a flat, though inclined, plane, dipping to the NNE at about 30°.

A similar construction allows structure contours for the top of the conglomérate to be constructed (diagram **A** continuous lines) and this dips at 36° to the SSE. As the base of the conglomerate in bore-holes **A**, **B** and **C** lies 50 m vertically below the top, structure contours for the base will lie along the same lines as those for the top, but are 50 m less in value (diagram **A**).

The various values of height for the geological formations imply that the sandstone overlies the conglomerate and the different attitudes of the geological surfaces suggest that the contact is disconformable (either a fault or an unconformity). Points on the map where structure and topographic contours of like value cross (i.e. points of likely exposure) are shown in diagram **A** as small squares and small circles, as are points of possible intersection of the base of the sandstone with the top and bottom of the conglomerate (large shaded squares). Note that these intersections lie on straight lines trending almost east to west.

By 'joining the dots', but being aware of the disconformable base of the sandstones, we can construct the outcrops of the various surfaces as in map **B**. Because the conglomerate is truncated by the sandstone its outcrop *on* the plane of contact will be a straight band (intersections of flat, though inclined, planes must be straight lines; see Fig. 33). Note that the outcrop of the conglomerate does not pass through some of the localities determined as potential outcrops because it was eroded away before deposition of the sandstone. As the conglomerate dips to the south, the outcrop of its base at the topographic surface, as well as with the unconformable base of the sandstone, defines the northern limit of ground underlain by the conglomerate (diagrams **B** and **C**).

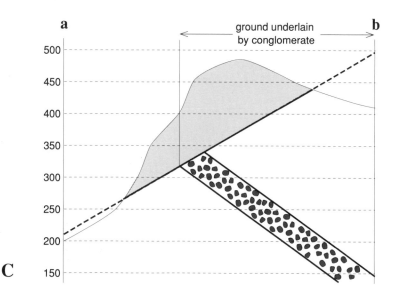

Exercise 7.2.1

Because it runs nearly parallel to the topographic contours, the dolerite–mudstone contact is flat-lying. Structure contours show that it is a flat plane dipping at a shallow angle to WSW (diagram **A**). In contrast the dolerite–sandstone contact is markedly curvi-planar with dips away from the centre of the map. Structure contours are consequently curved and define an irregular dome shape (diagram **A**). Note that in several places the dolerite dies out so that the mudstones and sandstones come into contact. The dolerite was intruded along the sandstone/mudstone contact by arching up the sandstones.

The vertical thickness of the dolerite can be determined by noting the differences in height between structure contours for its base (the mudstone/dolerite contact) and its top (the dolerite/sandstone contact). These points of known thickness are shown in diagram **B** as black and white circles. Knowing that the dolerite has a flat base and dome-shaped top, these data points can be contoured (as in diagram **C**) to give a map of the variation in thickness and also the original extent of the dolerite (diagram **D**).

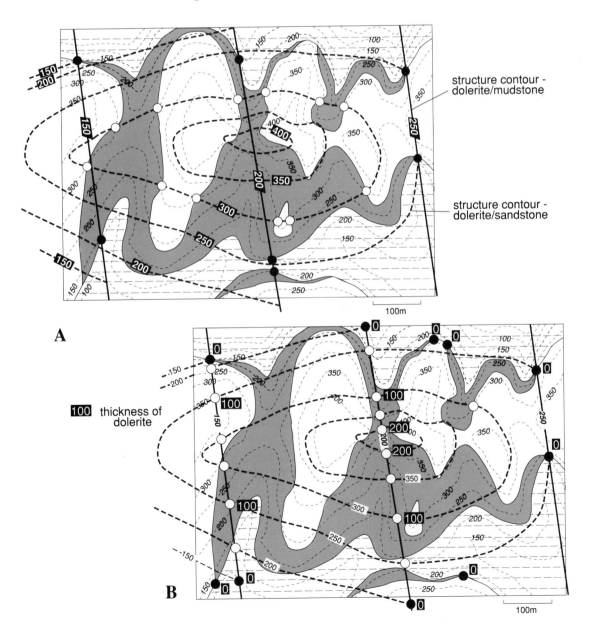

A

structure contour -
dolerite/mudstone

structure contour -
dolerite/sandstone

100m

B

100 thickness of
dolerite

100m

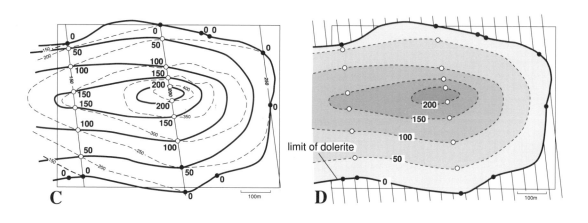

C

limit of dolerite

D

100m

100m

Exercise 7.2.2

In the south of the area the conglomerate clearly cuts across the contacts within the volcanic ash/lava/sandstone/mudstone sequence. It occupies the tops of the hills and ridges and therefore unconformably overlies the other rocks (diagram **A**).

The volcanic ash occurs in the valleys in both the north-west corner and the south of the map. It is succeeded upwards in turn by lava–sandstone– mudstone, suggesting that this may be the stratigraphic sequence. Structure contours for the base of the conglomerate show that the unconformity is an undulating surface dipping to the WSW. The attitude of the underlying rocks is not easy to assess but structure contours for the sandstone/mudstone contact reveal an asymmetric, synformal fold plunging to the WSW (diagram **B**). Thus the conglomerates truncate a major fold affecting all the older rocks.

In assessing the shape and thickness of the mudstone we can firstly construct isopachytes using the structure contours for the conglomerate and the mudstone/sandstone contact. Diagram **C** shows how the thicknesses are

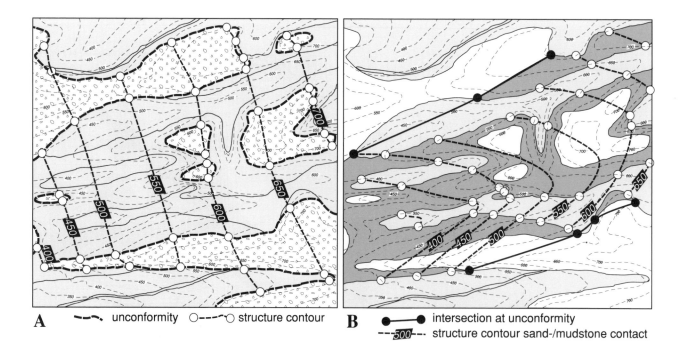

A ---◦--- unconformity ◦----◦ structure contour

B ●——● intersection at unconformity
---*500*--- structure contour sand-/mudstone contact

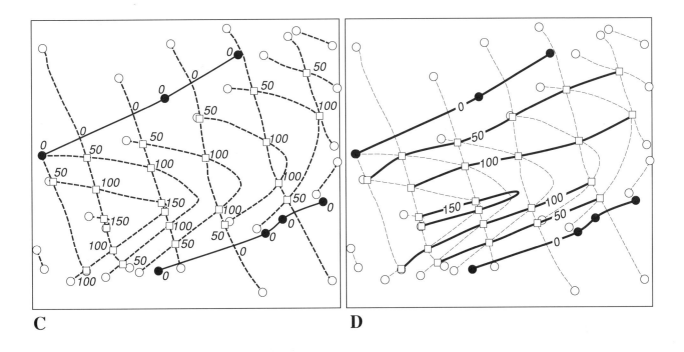

C   D

calculated from the intersections of the two sets of structure contours and **D** shows contours drawn for these values. Note in these diagrams and **B**, that the lines of intersection of the mudstone/sandstone contact with the unconformity (mudstone thickness = 0 m) are also constructed. Erosion prior to deposition of the conglomerate has removed the mudstone to the north-west and south-east of these two lines. In diagram **D** we can see the thickness variations for the mudstones as they would have been before further erosion produced the present land surface. In order to establish present thickness variations, we have to consider the effects of this erosion because, in places, valleys are cut through the mudstones to expose the underlying sandstone; the mudstone has been removed.

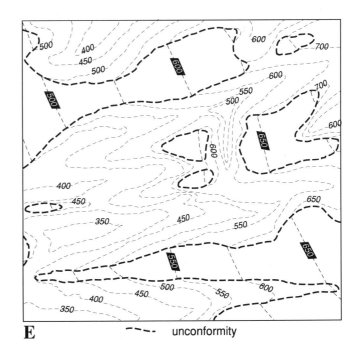

E   - - - unconformity

In diagram **E** the thickness of the mudstone is calculated for points where stucture contours cross topographic contours so that isopachytes for the present thickness can be constructed. Note that the outcrop of the sandstone/mudstone contact represents the zero thickness contour and where the mudstone underlies the conglomerate we use the isopachytes from diagram **D** to assess thickness. Thus we can predict that the maximum thickness of mudstone, over 100 m, occurs in the middle of the eastern half of the area (diagram F).

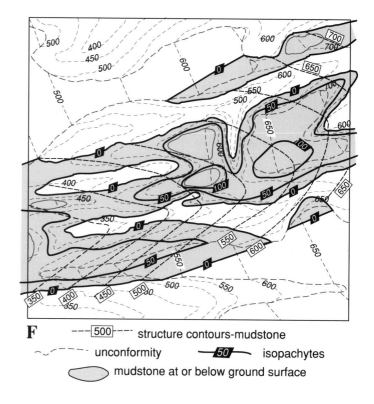

F ---[500]--- structure contours-mudstone

~~~~ unconformity   —[50]— isopachytes

[mudstone ellipse] mudstone at or below ground surface

Exercise 8.4.1

Assuming that before fault movement the outcrops of the basalt dyke were continuous, as were those of the sandstone/mudstone contact, the only intersections present that will allow us to determine slip on the fault are those between the basalt and the sandstone/mudstone contact.

The basalt intrusion is a vertical dyke (it has a straight-line outcrop across the topography), the fault dips to the SSE, as does the sandstone/mudstone contact on both sides of the fault (diagrams **B** and **C**). As the dyke is vertical, structure contours for its margins lie vertically above each other along the lines of the margins (diagram **E**). Consequently the lines of intersection between the dyke and the contact within the sediments will trend along the dyke margins (diagrams **D** and **E**). By noting the values for the structure contours of the contact along the dyke, we can determine the attitude of the intersection: south of the fault it plunges to the SE from 600 to 400 m whilst to the north it plunges in the same direction from 400 to less than 200 m (diagram **D**). The points along the line at 400 m coincide in position with the 400 m structure contour for the fault and therefore these are the positions where the line of intersection hits the fault plane. Before fault movement they were in the same place and therefore a line between them, on the fault plane, gives the slip direction (diagrams **D** and **E**). However this assumes that slip on the fault was unidirectional. Given that this was so, movement on the fault involved about 400 m of slip along the strike of the fault.

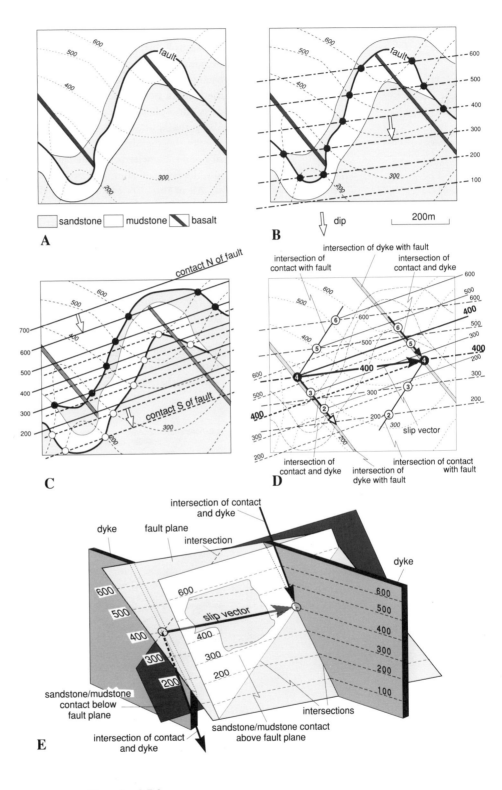

sandstone   mudstone   basalt

A

B

dip   200m

C

contact N of fault

contact S of fault

intersection of dyke with fault

intersection of
contact with fault

intersection of
contact and dyke

slip vector

intersection of
contact and dyke

intersection of
dyke with fault

intersection of contact
with fault

D

dyke   fault plane

intersection of contact
and dyke

intersection

dyke

slip vector

sandstone/mudstone
contact below
fault plane

intersection of contact
and dyke

sandstone/mudstone contact
above fault plane

intersections

E

Exercise 9.7.1

Looking at the relationships between the geology and topography, given that we know the order in which the sediments were laid down (the stratigraphic sequence), we can see that adjacent to the fault old rocks overlie younger, a situation that usually arises with reverse or thrust faulting. In the SW corner of the map the fault is flat-lying, at about 180 m. Immediately to the north, on the other side of the ridge, it is exposed again, at about 180 m, and then further to the north it drops in height to 100 m. The flat-lying fault cropping out to the east of this, in the next valley, lies at about 75 m and thus is likely to be the same fault (it has a similar attitude and lies at about the same height). This conclusion is supported by the occurrence of the same rock unit directly above the fault in both places. Thus whilst in the SW and NE it is flat-lying, the fault is curvi-planar dropping in height to the NE.

Above the fault, outcrop shape and evidence for the heights of contacts (white circles in the diagram) indicate dips, as shown by the arrows and marginal notes—likewise below the fault. In places (bold circles) contacts above and below the fault are truncated by it and points (and lines) of intersection are apparent. All of this data can be projected onto the line of section because the common NW–SE trend of the few structure contours that can be drawn, and the lines of intersection, show that the strike of the fault and of the contacts are parallel.

The section can now be drawn and reveals what appears to be a listric thrust fault with a roll-over anticline above it. If we assume that slip has been up the dip of the thrust ramp, then displacement on the fault plane would be distance **d** (i.e. the distance between the intersection points of the base, or top, of the sandstone with the fault plane). Before faulting, the cut-off of the base (or top) of the sandstone at **y** on the fault plane would have been at **x**. This gives a horizontal shortening of the crust as shown (the line of section is at right angles to the direction of strike and therefore in the dip direction of the ramp). Note, however, that the total displacement **d** is measured *along* the fault plane.

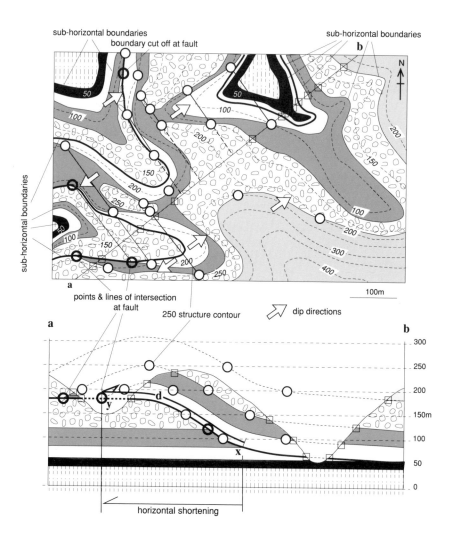

Exercise 9.7.2

Examination of outcrop shapes relative to topography reveals that, in the west, fault **a** dips at a moderate angle to the east, but traced eastwards it flattens out to become horizontal. Fault **e** again dips eastwards and merges with **a**. Mirror images of these faults occur on the southern side of the main valley (faults **f** and **g** respectively) so that fault **a** = **f** and **e** = **g**. Faults **b** (= **c**) and **d**, in contrast, dip westwards at steep to moderate angles and are truncated by fault **a**.

In the west and beneath fault **a** the sediments appear to be horizontal, whereas above the fault they dip westwards at shallow to moderate angles. Location of structure contours and intersections of planes (diagram **A**) confirm these deductions and allow a cross-section to be constructed. For the sake of clarity, only some of the data points relevant to the attitudes of the faults and the top of the limestone are shown in diagram **A**. Note that location of the lines of intersection of the sediment boundaries and the faults (square boxes) are equally as important as structure contours in determining the position of these boundaries in the section. The completed cross-section (diagram **B**) shows that listric extensional faults and associated antithetic faults cut the sediments and the dykes, and fault movement has caused the development of a roll-over anticline. Assuming dip-slip, amounts of movement *along* the fault planes are given by the displacements of various markers (half-arrows in diagram **B**). We can also measure the total horizontal extension

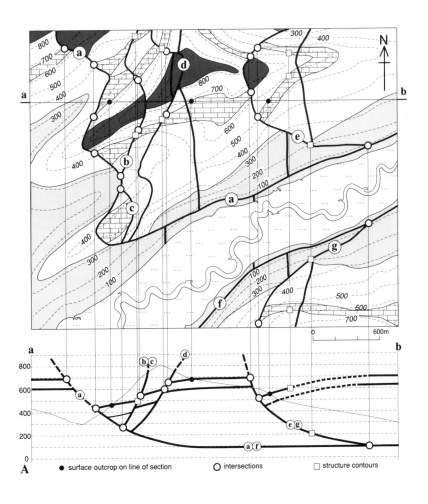

effected by the faults. Points **x** and **y** on the sandstone/mudstone contact now have a horizontal separation of l_n (≈ 3090 m) but before faulting they would have been separated by: $l_1 + l_2 + l_3 + l_4 + l_5$ as measured along the contact on the section (≈ 2870 m). Thus the extension is $3090 - 2870 = 220$ m.

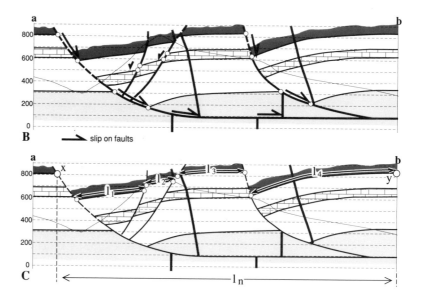

Exercise 10.3.1

Veeing of outcrops in valleys and over ridges indicates north-north-westerly and south-south-easterly dips (arrows in diagram **A**). Changes in dip suggest the presence of an antiform and synform. Curved outcrop shapes on valley- and hill-sides locate the positions of fold hinges (black circles). Except in the north, straight-line structure contours do not fit with the outcrop shapes which in most places are smoothly curving, not angular (diagram **A**). The data points do however fit elliptical structure contours (diagram **B**), reflecting the 'whale-back' shape of the antiform.

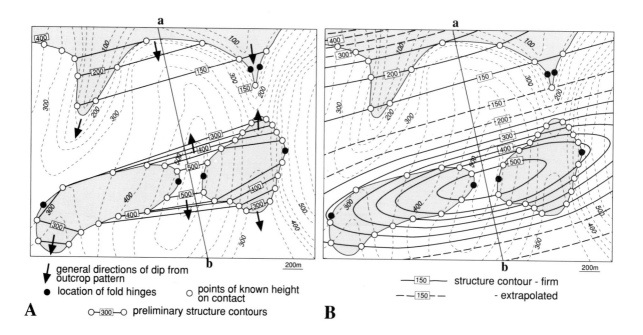

Note that structure contours drawn with a N–S trend would form a complex pattern and do not fit with the outcrop shapes. Whilst the synform is essentially cylindrical with a horizontal hinge, the antiform is non-cylindrical and its hinge plunges to the WSW at its western end, and to the ENE at its eastern end (diagrams **C** and **E**).

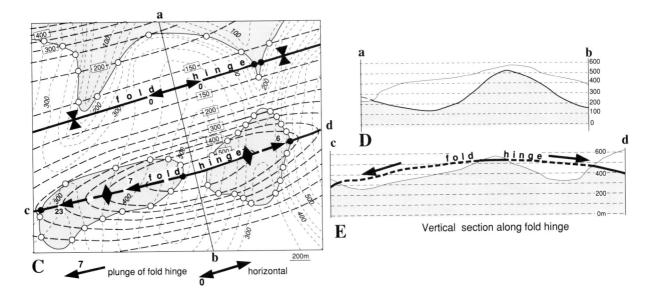

Exercise 11.2.1

Curvature of the outcrops on the two sides of the valley indicate the presence of an antiformal/synformal fold pair and mirror images of the outcrops across the valley suggest that these have essentially horizontal hinges and trend from north-east to south-west. The shared limb dips steeply to the east whereas the long limbs of the folds dip at low to moderate angles to the west. The folds are therefore asymmetric. The succession, in ascending order of height, is **a** to **b** to **c** to **d** to **e**; **a** crops out in the core of the antiform in the deepest parts of the valley; **e** occupies the high ground and cores the synform.

The shapes of outcrops around the fold closures suggest in places that these are rounded, in others more angular (diagram **A**); the fold limbs are straight. Construction of structure contours for the contacts (only those for the base of **b** are shown in diagram **B**) confirms the deductions made above and allows a cross-section to be drawn (diagram **C**). Places along contacts for which structure contours can be located are shown as thick lines, and it is important to note that the thicknesses (**t**) of the two sandstone formations on the fold limbs are constant. The upper contacts of the sandstone formations are rounded over the core of the antiform and angular in the core of the synform. Conversely the lower contacts of these formations are angular in the core of the antiform and rounded around the synform. These observations permit a reasonably accurate assessment of the geometry of the fold profiles to be made. Attaining such accurate profiles is important because knowledge of fold geometry affects the way in which we calculate the position of the oil-bearing sandstone.

Careful construction of the cross-section gives us the attitudes of the axial surfaces of the two folds which must, at depth, affect the oil-bearing formation. Because we know that the overlying mudstone is 220 m thick, and we know the fold geometry, we can determine the position of the top of the oil sandstone as in diagram **C**. We can therefore predict where bore-holes should be sited, i.e. along the crest of the antiform (diagram **D**). Location **A** is preferred because the rock overlying the oil sandstone (the overburden) is thinnest here.

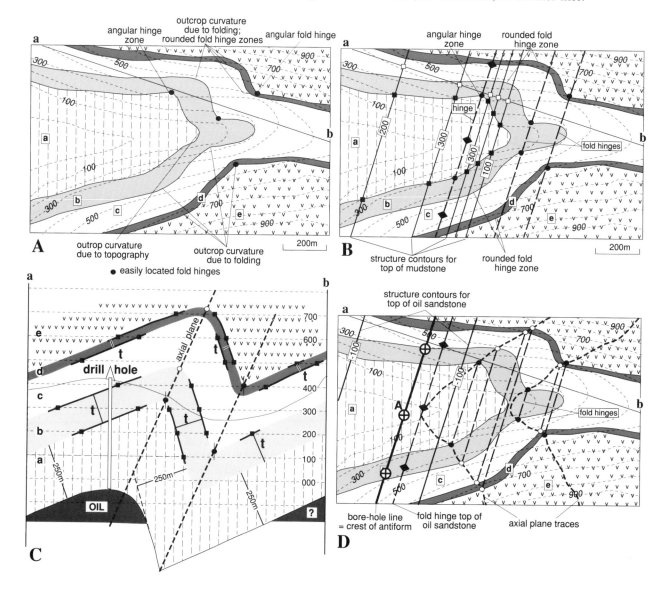

Exercise 11.6.1

The variations in dip, as indicated by the dip and strike symbols and the occurrence of curving outcrops unrelated to topography, show that the rocks are folded. The consistent strike further shows that the folds must be cylindrical with horizontal hinges. Note that the strike of the rocks is also given by the lines joining locations where junctions—between the volcanic rocks and the mudstones; the mudstones and the siltstones; and the conglomerates and the siltstones—lie at sea level (black circles in diagram **A**).

The possible positions of fold hinges are indicated by the white circles (i.e. places where rapid curvature of outcrops do not appear to be related to topography). These are confirmed as fold hinges because they lie on straight lines parallel to the strike of the beds (dashed lines). Bearing in mind that the outcrops of axial planes will run through points of emergence of fold hinges, will separate areas of different dips characteristic of fold limbs, and will change trend according to their dip and topography; the approximate axial plane traces can now be drawn in (thick dashed lines). Two flat-lying folds are present, both of which are overturned towards the east.

Because the folds are cylindrical, with horizontal hinges, the cross-section can now be drawn by extrapolating all of the data onto the line of section (diagram **B**). Note on the section, however, that we only know the heights of those localities for which spot heights are given (circles with dip lines) and where junctions lie at sea level (small black circles); we do not know the heights of the fold hinges although we do know their positions along the line of section (vertical dashed lines with open circles). The final cross-section (diagram **C**) is based on extrapolation of the position of contacts away from where they are firmly controlled (heavy lines) and knowledge of the positions of fold hinges along the line of section (white circles with dashed lines). As the section is drawn at right angles to the fold hinges, the thicknesses of the various formations are true thicknesses and thus we can see that the formations show thickening in the cores of the folds.

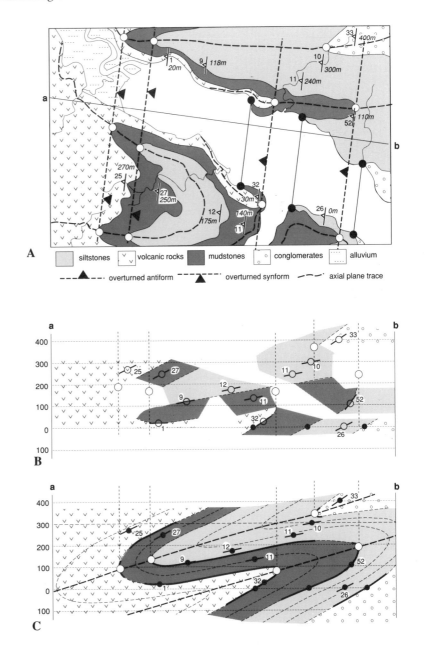

Exercise 11.6.2

By extrapolating outcrops of the boundaries between exposures and assessing the changes in strike, we can locate three major folds which, because of the dips of the bedding and the plunge of associated folds, must plunge to the north-east at about 11° (diagram **A**). As the directions and amounts of plunge of the folds are the same, the folds must be cylindrical and therefore we can, by projecting all the data on the map along plunge (as in Fig. 88), construct a cross-section. Because the topography in this case is flat and horizontal, lying at 400 m above

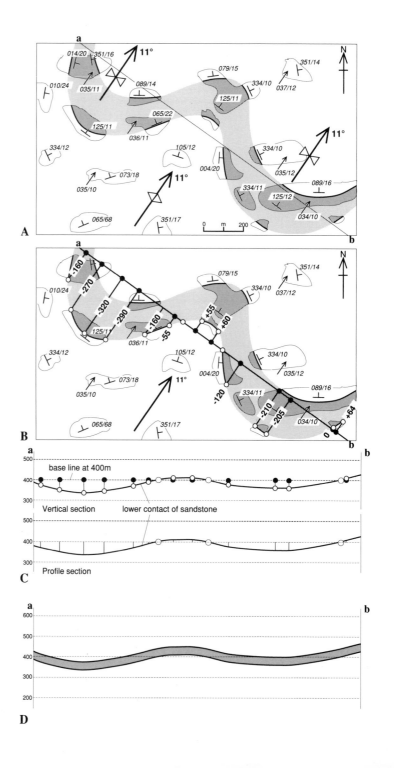

sea level, we can plot the data points selected (white circles in diagram B) by measuring on the map, in the direction of plunge, the distance from the section line. Negative values are for those points on the selected surface that will lie below ground surface (400 m) and positive values are for those lying above. In each case height above or below base level is given by the relationship $\tan p = h/d$, where p is the angle of fold plunge, h is the height above or below base level and d is the distance from the section line (see Fig. 88). When plotted, the data points delimit the folds as in diagram C. The full cross-section is shown in diagram D. Note the differences in the shape of the folds when seen on the map as compared with the section (cf. Fig. 79). This difference is due to the 'cut effect'; neither gives the true shape of the fold because neither the topographic surface nor the plane of section lie at right angles to the fold plunge. In order to construct a true profile section we would use the relationship $\sin p = h'/d$, where h' is the height above or below base level as measured in the profile plane at right angles to the fold hinge.

Exercise 12.4.1

At first glance the map presents a bewildering array of data—but don't panic! Provided you approach the problem systematically and apply the analytical methods outlined below, an accurate assessment of the structure can easily be gained. The exercise, though unrealistic only in as much as seldom would the strike of bedding and cleavage be so consistent, is one that reflects problems encountered during the mapping of folded rocks. Remember:

(i)   The outcrops of inclined geological surfaces, be they contacts, faults or axial surfaces, will 'vee' in relation to topography and dip.

(ii)   Fold limbs and hinge zones can normally be identified by systematic differences in the dip of the layering and/or changes in the relationships of cleavage to bedding dip.

(iii)   Hinge zones are characterised by unusually low or high dips of the layering and by cleavage lying at a high angle to the layering.

In this case the dips of the layering (bedding) are, with few exceptions, towards the NE at moderate to low angles, and areas of relatively lower and steeper dips can be distinguished (diagram A). This suggests that the area contains overturned folds which, because of the consistency in the strike of both bedding and cleavage, must be cylindrical with horizontal hinges. The shaded and blank areas in diagram A thus represent the approximate outcrops of alternate limbs of major folds. This is confirmed by further noting exposures where cleavage dips more steeply than bedding (heavy open circles with and without a cross line)—normal limbs; where it is less steep (black circles)—overturned limbs; and where there is a high angle between bedding and cleavage (heavy circles with cross)—hinge zones. Thus the boundaries between the shaded areas represent the outcrops of the axial surfaces of the major folds and we can now draw in the outcrops of the grey/green slate contact (diagram B). In doing this we note the exposures where the trend of the contact is given and take into account both the 'veeing' of outcrops in relation to topography and the positions of the fold hinge zones (the outcrops of the axial surfaces).

As we know that the fold hinges are horizontal and the folds cylindrical, hinge lines will trend parallel to the strike and connect repetitions of the outcrop of the fold noses across valleys and hills (diagram **B**). Note in the cross-section (diagram **C**) that the cleavage fans around the folds.

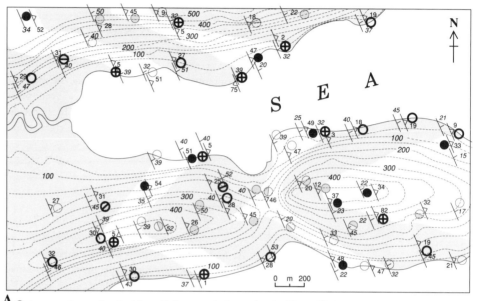

A   ◯ cleavage steeper than bedding   ● cleavage shallower than bedding   ⊕ cleavage at high angle to bedding

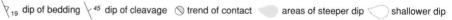

⊽₁₉ dip of bedding   ⟍⁴⁵ dip of cleavage   ◌ trend of contact   ▒ areas of steeper dip   ◌ shallower dip

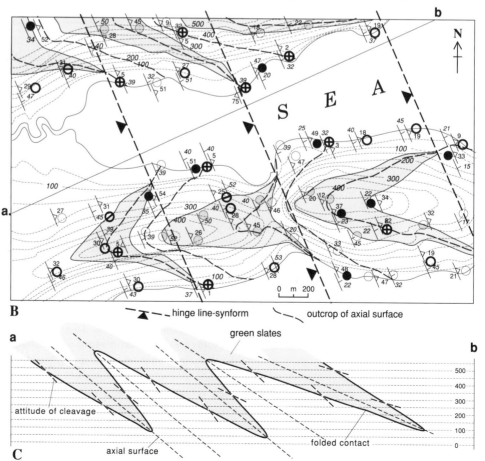

B   ◤---- hinge line-synform   ⌐⌐ outcrop of axial surface

Exercise 13.1.1

The shift of the outcrops of the limbs of the synform across the fault suggests that the block to the north has been downthrown relative to the south. This because outcrops of the light grey formation to the north are on the limbs of the fold but to the south they are closer to the fold core; the width of outcrop of the fold (**w**) changes. There also appears to be a horizontal offset; the fold hinges in the southern block have been shifted slightly to the west (diagram **A**). On this evidence alone it is not possible to determine the true slip direction on the fault; it may involve horizontal or oblique-slip or it may not.

Location of the same fold hinge on both sides of the fault (that for the base of the upper, dark grey formation) shows however that to the north it lies at 500 m whereas to the south of the fault it is at 650 m. Consequently movement on the fault must involve a component of vertical displacement. Projection of the fold hinge onto the fault plane shows (diagram **B**), that if movement on the fault was unidirectional, it was in its dip direction. The fault is therefore a normal fault with no horizontal component of slip, though there is horizontal shift. The latter arises because the strike of the fault plane is oblique to the trend of the fold hinges.

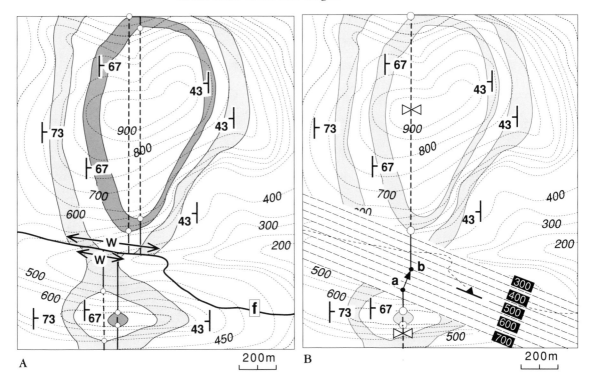

Exercise 13.2.1

Examination of the shift of outcrops across the faults (diagram **A**) does not allow us to determine the slip directions on the faults—amounts of shift are not consistent and evidently we need more data. Examination of outcrop shapes relative to topography reveals general dip directions (arrowheads in diagram **B**) and the presence of fold hinges (circles). The two faults must be vertical as they do not change trend with changes in topographic height. Three folds are present in fault block **a** and four in blocks **b** and **c** (diagram **C**). The sequence of

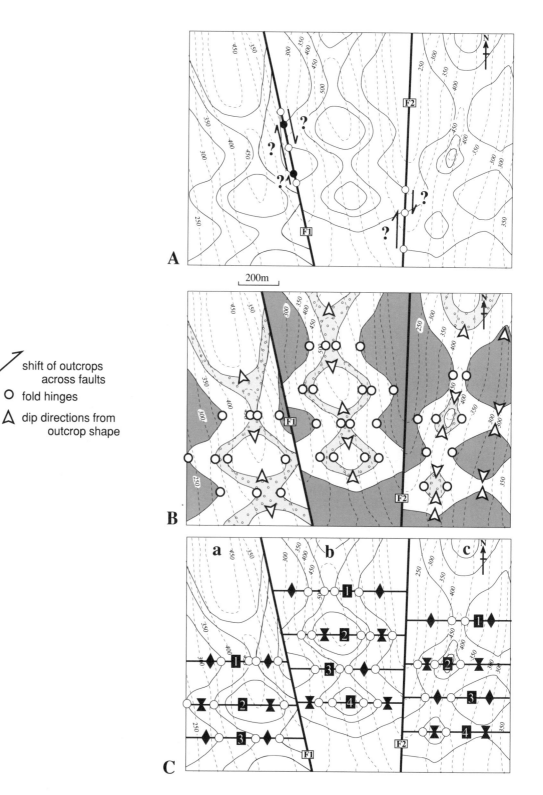

200m

shift of outcrops
across faults

fold hinges

dip directions from
outcrop shape

folds (antiform–synform, etc.) and the horizontal spacing of the axes suggests that the folds can be correlated between the fault blocks so that fold 1 is the same in each, and so on. Construction of N–S cross-sections through each fault block confirms this (diagram **E**). The fold hinges are horizontal so that their heights can easily be compared in order to assess the slip on each fault. Note that for fold 1 the fold hinge for the siltstone/mudstone contact lies at about 375 m in block **a**, at 425 m in **b**, and 425 m in **c** (diagram **D**). Slip on fault 1 therefore involves a horizontal component and a 50 m downthrow on its western side; it appears to be an oblique-slip fault. Fault 2 involves no vertical

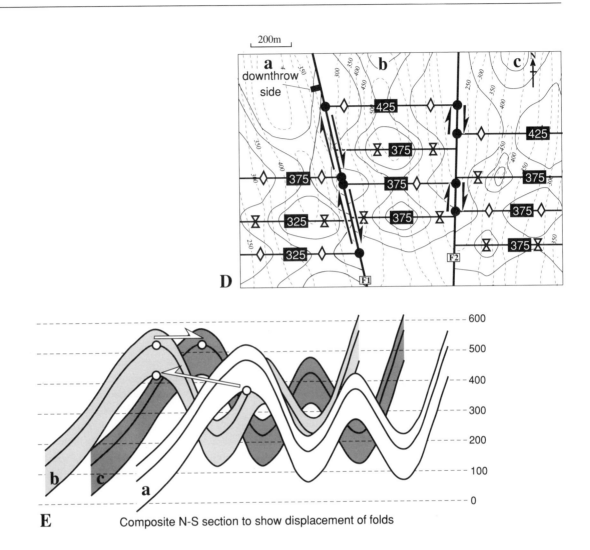

E   Composite N-S section to show displacement of folds

movement but horizontal slip has displaced the fold hinge; it appears to be a strike-slip fault. By checking the displacements of the other fold hinges we can confirm these conclusions and show that there is no variation in slip along each of the faults, e.g. there are no rotational components to the slip. Slip on the faults is illustrated in the composite N–S cross-section (diagram **E**).

Exercise 13.2.2

For ease of description the folds and faults have been numbered and the different fault blocks are identified by letters A to E (map **A**).

The general trend of the outcrops of the sedimentary rocks and the strike symbols tell us that the rocks consistently strike just west of north. 'Veeing' of outcrops in the river valleys and the dip symbols show the rocks to be folded and reveal that the siltstone cores antiformal folds, the conglomerate synforms. Because the outcrops of the fold cores do not close to north or south, the fold axes must be essentially horizontal. The overall consistency in widths of outcrops along strike *within* each fault block indicates the cylindricity of the folds. Similarities in outcrop shape across the river valleys suggest similar angles of dip and therefore that the folds are essentially upright and symmetrical. Bearing these observations in mind the fold axes can be drawn on the map as in diagram **A**.

The outcrops of faults 1 and 4 are straight: the faults are therefore vertical. The overall NE–SW trend and 'veeing' across valleys suggest that faults 2 and 3 strike NE–SW and dip at moderate to steep angles to NW and SE respectively.

Examination of the pattern of folding across the map reveals that the three folds in block A have approximately the same outcrop widths and spacing between their axes as folds 2, 3 and 4 in block B. The same is true for the folds in block D when compared with those of A and B. Despite differences in outcrop width it is also noteworthy that the spacing of the axes of folds 1, 2 and 3 of block C appear to match 1, 2 and 3 of B. Likewise folds 3 and 4 of block E match 3 and 4 of blocks B and D. These similarities strongly suggest that folds 2, 3 and 4 are respectively the same in each of the fault blocks. Therefore they can be used to analyse fault displacements.

Superficial appraisal of the shift of fold axes across the faults might suggest left-lateral (sinistral) strike-slip movement on each fault, i.e. the blocks to the immediate north of each fault have moved to the left relative to the blocks to the south. However, Chapters 8 and 13 of this manual have indicated that more rigorous analysis is usually necessary to determine slip directions.

Similarities in the widths of the outcrops of the cores of folds 2, 3 and 4 between blocks A, B and D have already been noted. The situation is therefore similar to that illustrated in Fig. 103 (diagrams **D** to **F**) and implies that movement between these blocks was essentially horizontal, i.e. strike-slip. In contrast, changes in the widths of outcrops occur between blocks A – C, C – D, D – E and E – B, indicating vertical components of movement (cf. Fig. 103, diagrams **A** to **C**). West of fault 2 the width of the synformal fold core (fold 2) increases whilst that of the antiform (fold 3) narrows. Thus fault 2 must downthrow on its western side. Similar analysis of fault 3 reveals a downthrow on its eastern side. In both cases the downthrow is therefore on the side towards which the faults dip. Displacement could however be either dip or oblique-slip.

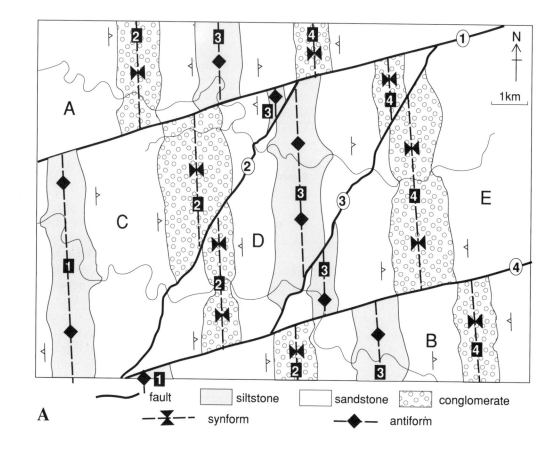

fault   siltstone   sandstone   conglomerate

synform   antiform

A

In map **B** the horizontal slip on faults 1 and 4 between blocks A – D and D – B respectively is shown by the half-arrows connecting the axes of folds 4 (fault 1) and 2 (fault 4). In each case it is about 2000 m. Because the similarities in the widths of fold cores across blocks A, D and B preclude vertical components of displacement, the faults must be strike-slip faults. However, the horizontal shift of folds 2 and 3 across fault 1 between blocks A and C is less than this, i.e. 1500 m. The same amount of shift is also apparent between blocks B and D (across fault 4) for folds 3 and 4. These differences arise because of the effects of faults 2 and 3.

Because the downthrow on fault 2 is on its western, dip, side, it must be extensional and its direction of slip is, in general terms, similar to that on fault 1. Both blocks C and A have moved towards the WSW but displacement of C has been achieved by slip on both faults whereas block A has only moved along fault 1. Similar arguments apply to faults 3 and 4 though movement is in the opposite direction.

It follows from this analysis that, along fault 1, slip was essentially horizontal between blocks A and D but, between A and C, and A and E, it involved both vertical and horizontal components. A similar situation is true for fault 4 (diagram **B**).

Because faults 2 and 3 do not occur in blocks A and B—they are neither displaced by, nor do they cut faults 1 and 4—they must have formed at the same time as movement on faults 1 and 4. They have allowed extension and subsidence in the strike-slip zone lying between faults 1 and 4 (cf. Figs. 58 and 59). The slip directions on faults 2 and 3 are therefore likely to be oblique and in the direction of strike of the controlling major strike-slip faults (arrows in map **B**). The shortfall of 500 m horizontal slip between blocks A and C, and E and B, is accommodated by oblique slip on faults 2 and 3. The situation is therefore similar to that illustrated in Fig. 58.

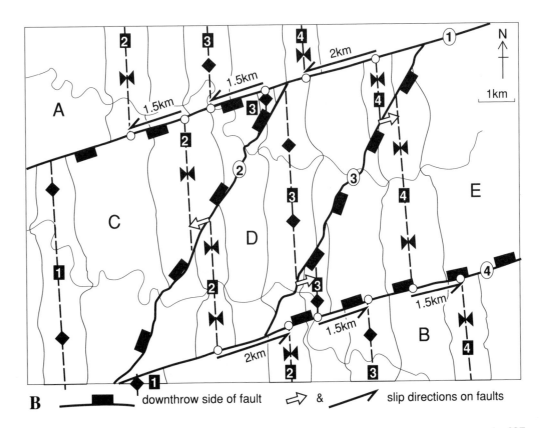

Exercise 14.0.1

Outcrop shapes in relation to topography and repetiton of sedimentary formations reveal that gently dipping faults cause repetition of the stratigraphic succession; note particularly how the conglomerate is repeated in the packages of rock sandwiched between the various faults. It is also apparent that the pattern of outcrops on the hillsides across the main valley are mirror images of each other, showing that the structures trend from NW to SE. Whilst many of the faults truncate, or join, each other, we can recognise, in places, three main faults, labelled **a**, **b** and **c** in diagram **A**. Outcrop shape in relation to topography tells us that in the west and the east of the area, fault **a** is horizontal but changes in height from about 350 to 150 m. The drop in height takes place in the middle of the map area with a gradual change of attitude. The fault is therefore listric (refer back to Chapter 9). In the west of the area both faults **b** and **c** dip to the west; in the east they dip to the east. Judging from their topographic heights, fault **c** overlies **b** which overlies **a**. Fault **c** would also appear to merge with faults **a** and **b** in the west, as do both **b** and **c** with **a**, in

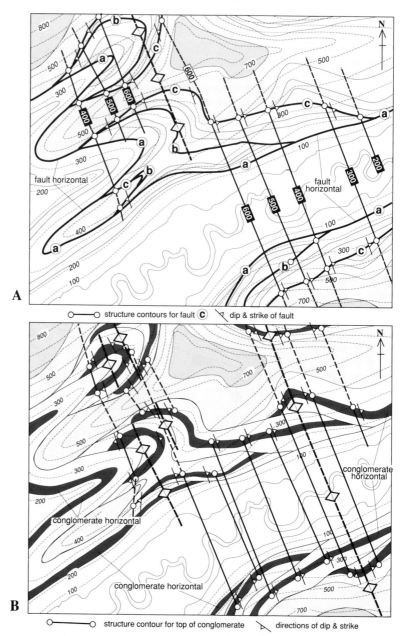

the east (diagram **A**).

Structure contours for fault **c** are shown in diagram **A** and demonstrate its consistent NW–SE strike but variable dip. The fault is curved, with the hinge of a pronounced antiform trending NW–SE across the area (diagram **A**). Construction of structure contours for the other faults (not shown in diagram **A**) shows that these also have curving trajectories.

Consideration of outcrop shapes and structure contours for the non-faulted contacts between the various sedimentary formations reveals that they are again folded (e.g. the top of the conglomerate—diagram **B**). Note that whilst the antiform in the east of the area extends across the major valley it does not affect the rocks below fault **c**. Likewise the other folds are truncated downwards by faults. Full analysis of the attitudes of the faults and the formation boundaries reveals a number of folds. By plotting their axial plane traces (diagram **C**) we can see more clearly that some of these folds are truncated by the faults whilst others fold them.

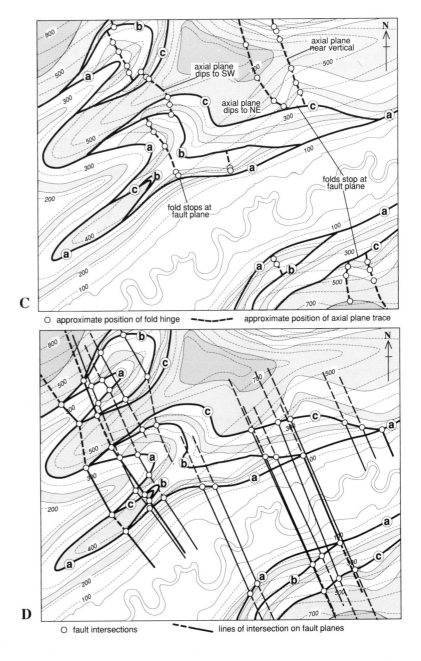

C

○ approximate position of fold hinge   – – – – – approximate position of axial plane trace

D

○ fault intersections   – – – – lines of intersection on fault planes

Note that in some places on the map the faults run parallel to the formation boundaries whilst in others they cut through the stratigraphic succession. In a number of places faults are parallel with the boundaries in the sediments above the fault plane (the hanging-wall) but discordant with those in the sediments below (the foot-wall). In other places this situation is reversed. In every case, however, the faults somewhere cut up through the stratigraphic sequence to the west. With the exception of parts of faults **a** and **c**, the faults emplace older sediments over younger and thus appear to be thrust faults. We can interpret their shapes and relationships to the stratigraphy in terms of ramps, where they cut up through the stratigraphy, and flats, where they are concordant (refer back to Chapter 9).

In constructing a cross-section to illustrate the structure, we make use of all the information above together with projections of lines of intersection between the faults and formation boundaries (diagram **D**). As examples, some of these data points are shown in diagram **E**, wherein the positions of the faults and boundaries are determined by the locations of structure contours and projected lines of intersection. In the section we can recognise a relationship between the location of the folds and the position of foot-wall ramps such that the former always occur up the dip of the faults to the west of these ramps (diagram **F**). The folds are roll-over anticlines (refer back to Chapter 9).

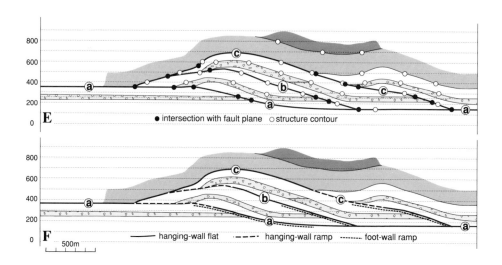

If we assume that displacements on the faults were at right angles to their strike, i.e. up the dip of the ramps, the section, being approximately normal to the strike, allows us to calculate the amount of slip that has taken place. Thus in diagram **G** the base of the conglomerate has been moved along the faults from a' to a″, b' to b″ and from c' to c″. We can measure these distances *along* the faults in the section (diagram **H**) and find that there has been about 575 m slip on fault **a**, 1250 m on **b** and 725 m on **c**. Remember, however, that in these calculations we have no direct evidence for dip-slip movement; there are no displaced markers that allow us to prove this.

If we assume dip-slip on the faults we can go a stage further in our analysis of the structure. From the map we can establish that the stratigraphic succession is the same throughout the area and that there are, at most, only minor variations in the thicknesses of the sedimentary formations. Thus before faulting there would have been a simple, uniform, layered sequence which was presumably horizontal. Because we know the amounts of slip on the faults and the uniformity of the stratigraphy, we can restore the section to its pre-fault configuration. To do this we choose a 'pin position' in the unfaulted area to the west (diagram **H**). From this we can measure the lengths of section along a chosen marker layer (here the base of the conglomerate), i.e. lengths **a**, **a–b** and **b–c** (diagram **H**). To do this you can use a length of thread laid along the appropriate line or pivot the edge of a piece of paper along it.

Knowing that the base of the conglomerate was originally horizontal, these lengths can be added to each other horizontally to take out the effects of the folding and fault movement, and to establish the original positions of the three faults (diagram **I**). Thus we can locate the original positions of the three faults where they cut the conglomerate. As from the map and section we can see that (i) each fault only penetrates down in the stratigraphic succession as far as the lowest siltstone and up as far as the upper siltstone/sandstone boundary, and (ii) the faults all merge upwards and downwards, we can draw in their trajectories on the restored section (dashed lines in diagram **I**).

The original horizontal length of the section is calculated by summing the bed lengths **a**, **a–b**, **b–c** and **c**, i.e. 6160 m (l_o in diagram **I**). After thrusting, this bed length was reduced to 3650 m (l_n in diagram **H**) so that the total horizontal shortening of the section, due to thrusting and folding, is given by $l_o - l_n$, i.e. approximately $6160 - 3650 = 2510$ m. Note that summing the horizontal distances l_a, l_b and l_c (diagram **H**), which equals 2400 m, does not give the true slip because it does not account for that element of shortening due to folding.

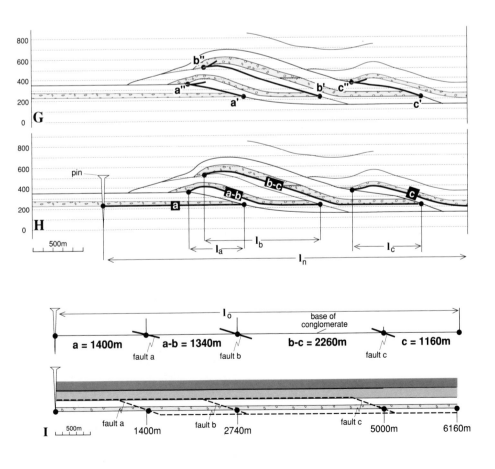

Exercise 14.0.2

The stratigraphic column gives the relative ages of the formations cropping out in the map area. From the column and the map we can see that the sandstones **a** and **b** appear to be conformable, with sandstone **a** restricted to the area to the SW of fault 4. Differences in dip and/or strike show that sandstone **b** may lie unconformably on the siltstone in the north and on the conglomerate and mudstone to the southwest of fault 4 (diagram **A**). Between faults 2 and 4, and 3 and 4, the sandstone and conglomerate appear to be conformable. A lateral change in rock type (facies) occurs across the area, from siltstone in the north to conglomerate in the south. Though these are different types of sediment, the stratigraphic column tells us that they are of the same age. They rest unconformably on mudstone which, in turn, unconformably overlies slate (diagram **A**). The conglomerate only occurs SW of faults 2, 3 and 4, the siltstone only to the NE of 2 and 3.

The volcanic rocks are restricted to the area NW of fault 1. Though intermediate in age between the conglomerate/siltstone and mudstone formations, they are absent from the area to the SE of the fault, both at surface and in the drill-holes. Either they were eroded away before deposition of the conglomerate/siltstone or were never deposited. The oldest rocks, the slates, form a basement to the younger sedimentary sequences and were deformed before deposition of the mudstone; none of the younger sediments or the volcanic rocks are cleaved.

Fault 1, because of its straight outcrop across topographic features, appears to be vertical. Fault 2, where it trends NW–SE, 'vees' in river valleys, indicating a dip to the SW. Likewise, where they trend NW–SE, faults 3 and 4 appear to dip to the SW. However, where these faults run NE–SW, they are vertical or nearly so. Because these changes of attitude occur without abrupt changes in outcrop trend or truncations, the faults must be scoop-shaped in three dimensions with their directions of strike swinging from NE–SW to NW–SE.

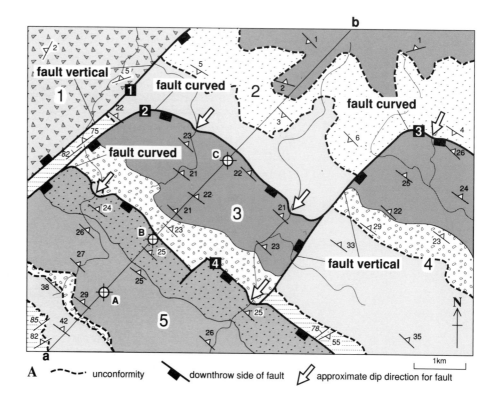

A — – – – unconformity   ◆ downthrow side of fault   ⬦ approximate dip direction for fault

Fault 1 appears to downthrow to the SE in the northern part of its outcrop but to the NW in the south-west (diagram **A**). Faults 2, 3 and 4 consistently downthrow to the south-west where they trend NW–SE, suggesting (but not proving) down-dip slip. Except for sub-areas 3 and 4, note that faults 2, 3 and 4, rather than cross-cutting each other, are linked together. Note also that the facies change from conglomerate to siltstone occurs across faults 2 and 3, indicating a connection between faulting and sedimentation.

On the basis of stratigraphy, dips and strikes, and the positions of the faults, we can divide the area into five structural/stratigraphic sub-areas:

Sub-area 1 lies to the NW of fault 1 and comprises only gently dipping volcanic ashes.

Sub-area 2 is bounded by faults 1, 2 and 3 and consists of a basement of slate unconformably overlain by mudstone dipping to the NE at a shallow angle. The succeeding unconformable siltstone

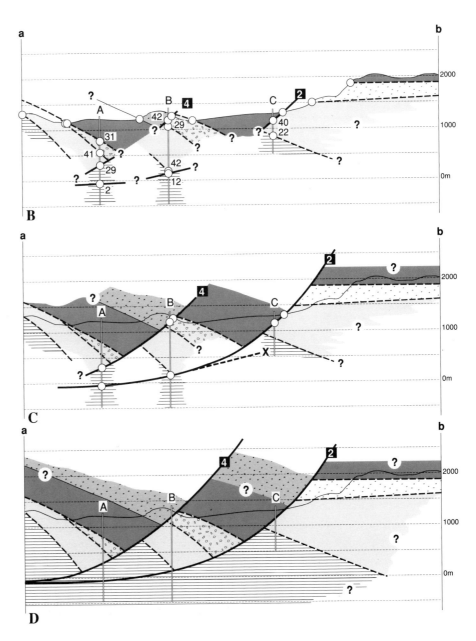

dips very gently to the SW and is overlain, probably unconformably, by sub-horizontal sandstone.

Sub-areas 3 and 4, bounded by faults 2, 3 and 4, are similar to each other. The apparently conformable conglomerate and sandstone lie above the mudstone and dip consistently to the NE at gentle to moderate angles. Note, however, in sub-area 4, the steeper dips in the mudstone and the occurrence of slates lying unconformably beneath.

Sub-area 5 contains the only outcrops of sandstone **a**: it appears to overlie conformably sandstone **b**. Dips in the uncleaved sedimentary formations increase progressively from NE to SW, i.e. away from fault 4. The basement is again exposed and note that the strike of the bedding and cleavage in the slate here is similar to that in sub-areas 2 and 4; there have been no dramatic rotations of rocks within the different sub-areas during faulting.

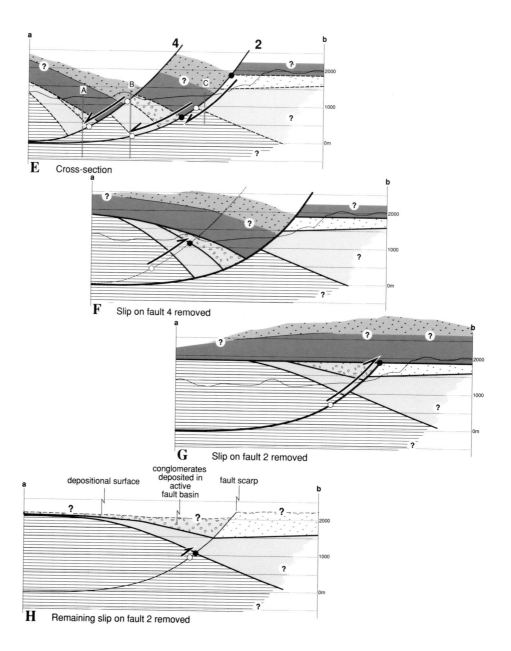

E   Cross-section

F   Slip on fault 4 removed

G   Slip on fault 2 removed

H   Remaining slip on fault 2 removed

From these observations and consideration of the discussions given in Chapter 9, it would appear that, with the exception of fault 1, we are dealing with a system of linked faults which may have influenced sedimentation. Though there is no direct evidence for directions of slip, consistency in downthrow relative to fault dip suggests extensional faulting. Whilst many of the rock sequences have been tilted from their original horizontal attitude, none, other than in part the slate, is overturned. Repetitions of succession relating to changes of dip are not present and therefore there are no indications on the map, other than in the slate, of substantial major folds. However, repetitions of succession do occur in relation to the faults.

The drill-hole data provide us with important information concerning the sub-surface structure. The upper part of the drill-hole log **A** matches the surface geology in that, at depth, the conglomerate/sandstone and conglomerate/mudstone contacts continue to dip to the NE. However, the mudstone is in fault, rather than unconformable, contact with the slate, and a second fault lies within the slate (diagram **B**). Because these faults dip to the SW, one of them, projected upwards, should connect with fault 4. As there are no faults at surface *within* sub-area 5, the upper one appears to be the most likely candidate (diagram **B**). We have to be cautious, however, as this could be a small synthetic fault branching off the lower one but not reaching surface.

The upper fault in log **B** is, because of similarities in dip direction and proximity, clearly the same as fault 4 and we can see that sub-area 3, like 2, 4 and 5, is underlain by a basement of slate (diagram **B**). The lower fault dips gently to the SW and therefore can probably be projected up-dip, to join fault 2. In log **C** clearly the fault is the same as fault 2, occurring at an intermediate height between its surface outcrop and its possible level in drill-hole **B**. This projection of fault 2 does however require a change of dip angle from about 40° at and near surface, to 12° at depth—the fault would be listric. If this is the case then it follows that the lower fault in drill-hole **A** could also be a continuation of fault 2; a ramp-flat geometry is implied. We can test these hypotheses by constructing the cross-section.

Diagram **B** shows an initial construction using surface data, dips and the drill-hole logs. Diagram **C** assumes that faults 2 and 4, as their dips suggest, connect, at depth, with the two faults in drill-holes **A** and **B**. Diagram **D** shows the completed cross-section and this suggests that the tilt of the sediments in sub-areas 3 and 5 is due to the development of roll-over folds relating to movement on the extensional, listic faults 2 and 4, sub-area 2 being a stable block.

Assuming down-dip slip on the faults and that we have drawn the cross-section as accurately as possible, we can see (diagrams **E** and **F**) that the amounts of offset of the various contacts across fault 4 are very similar. Offsets across fault 2 vary, however, with a smaller displacement of the base of sandstone **a** than of the base of the mudstone (diagram **E**).

By progressively removing the slip on faults 2 and 4 (diagrams **F** to **H**) we can see that, at the time of deposition of the conglomerate and siltstone, fault 2 was probably active, deposition being concentrated in the basin developing above the roll-over anticline (diagram **H**).

Faults 2, 3 and 4 can therefore be interpreted as a linked, partly syn-depositional, extensional fault system with faults 2 and 3 developing at the same time. Fault 4 would appear to be a slightly later but synthetic extensional fault.

We can say little about the nature of fault 1 except that it is not directly connected to the linked fault system. The absence of the volcanic ashes to the SE indicates either that the ashes bank up against it, i.e. it was present as a basin boundary during formation of the ashes, or that it involved a strike-slip component of movement bringing in rocks belonging to a different depositional regime from that preserved in sub-areas 2, 3, 4 and 5.

Exercise 14.0.3

From the relationships between outcrop shape and topography, dips and structure contours, it can be seen (diagram **A**) that:

(i)   The conglomerate occupies the high ground with its base lying horizontally at about 650 m. Because it does not change height across the area, it is unlikely to be affected by the faults; before erosion of the present topography it probably extended across the whole area. The conglomerate overlies the vertical dykes

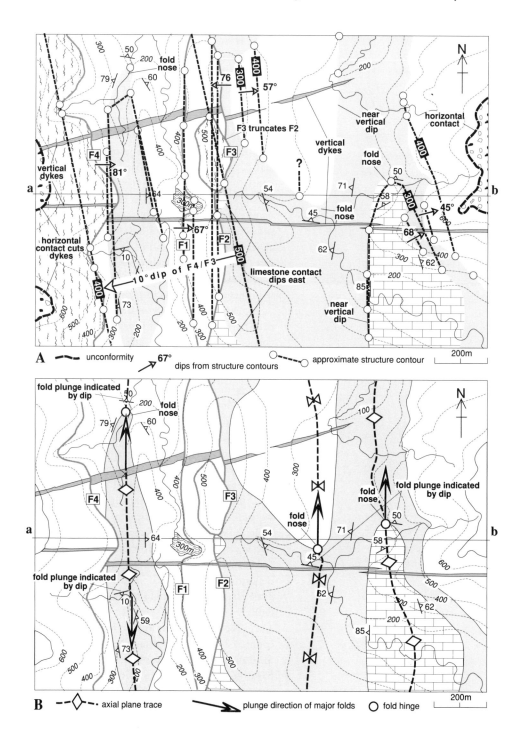

which cut the gneisses and the folded sediments. Therefore it seems to overlie unconformably all other rocks and the structures that affect them.

(ii)   The gneiss (a metamorphic rock that once lay deep in the Earth's crust) rests on faults **F3** and **F4** which have similar attitudes. Consideration of the respective heights of the faults and combined structure contours suggest that they are the same fault dipping to the WSW at 10°. As they previously lay at depth in the Earth's crust, the gneisses have therefore been uplifted along the fault.

(iii)   Fault **F1** dips to the east at about 67° and displaces the dykes, as does **F2**; **F2** however dips to the west at about 76°.

(iv)   The faults and dykes cut a group of sediments which, from west to east, are folded into an asymmetric antiform–synform–antiform (diagram **B**); the folds are not cylindrical, they form elongate domes and basins plunging both north and south; they are cut by the dykes.

(v)   In the west the antiform is cored by sandstone which is overlain by mudstone; in the east the antiform is cored by limestone overlain by sandstone; the intervening synform contains mudstone overlying sandstone. We can deduce from this a vertical sequence for the sediments: limestone (lowest)–sandstone–mudstone (highest) which is likely to be a stratigraphic succession. In turn these rocks are unconformably overlain by the conglomerate.

Thus a lower group of folded sediments is intruded by vertical dykes and cut by steeply dipping faults **F1** and **F2**. A younger, shallowly dipping fault

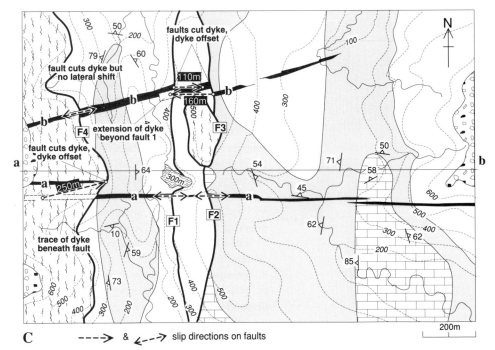

C   ┄┄➤ & ┄┄➤ slip directions on faults

F3–F4 emplaces gneisses above the folded and faulted sequence and all of these lower elements are unconformably overlain by the conglomerate.

We can now examine displacements on the faults in more detail. Fault **F1** displaces both dykes but note that dyke **a** is not shifted horizontally though it changes thickness. It trends at right angles to the strike of the fault and therefore the slip direction on the fault (diagram **C**) must be up- or down-dip (refer back to Fig. 48). It must be down-dip to the east because of the shift of dyke **b**. **F2** again does not shift the outcrop of dyke **a** horizontally and because of the offset of dyke **b**, slip must be again down-dip but to the west (diagram **C**). Both faults are therefore extensional, normal faults. **F3–F4** does not shift the outcrop of dyke **b** horizontally and thus, again, it must be a dip-slip fault; the shift of dyke **a** shows it to be a thrust fault bringing the gneisses up towards the ENE (diagram **C**).

Having determined the movement directions on the faults, we can now measure the amounts of slip by using the offsetting of the dykes. Dyke **b** has been shifted 110 m horizontally to the east by **F1** and 160 m to the west by **F2**. Dyke **a** has been moved 250 m to the ENE by fault **F4** (diagram **C**). Because we know the dip of the faults, the amount of slip is given by the relationship $\cos x = d/s$, where **x** is the angle of dip of the fault, **d** is the horizontal shift of the dyke in the direction of slip, and **s** is the amount of slip on the fault plane. For F3–F4, $x = 10°$, $d = 250$ m and therefore $s = 254$ m (diagram **D**). Note that by moving the hanging-wall above **F4** back by 250 m (the amount of displacement of the dykes), the gneisses still overlie the folded rocks (diagram **E**). **F4** must therefore, have been initiated before dyke intrusion, to be subsequently reactivated. Similar analysis of faults **F1** and **F2** reveals down-dip slip of 281 and 645 m respectively (diagram **G**).

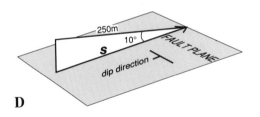

D

It is important to note that, without the evidence provided by the displacement of the dykes, we could say little about the nature of the faults. **F3–F4** places metamorphic rocks on top of sediments which have not been taken to depth in the crust and thus the gneisses must have been uplifted along **F3–F4**. Because of its dip, the fault is most likely to be a thrust, but we would not know its direction of slip. **F1** has the same mudstone on either side so that it could be a dip-, strike- or oblique-slip fault. **F2** causes, in the south, the younger mudstones to lie against the older limestones; the downthrow side is therefore to the west, but this situation could arise by dip-, strike- or oblique-slip (refer back to Chapter 8).

In constructing the cross-section (diagram **E**), use is made of the data recorded along the line of section, including measurements of dip, and some projected intersections. However, structure contours for the folded rocks are not used because they cannot be accurately located. In cross-section **F** the geological relationships are shown as they existed before intrusion of the dykes; note that the gneisses still overlie the folded sediments. Diagram **G** restores the section to its position before formation of the thrust fault (**F3–F4**) and diagram **H** illustrates the structure prior to movement of fault **F2**.

From the map we cannot determine the relationships between **F1** and **F2**, i.e. which cuts the other, but, given the extensional nature of both faults, and

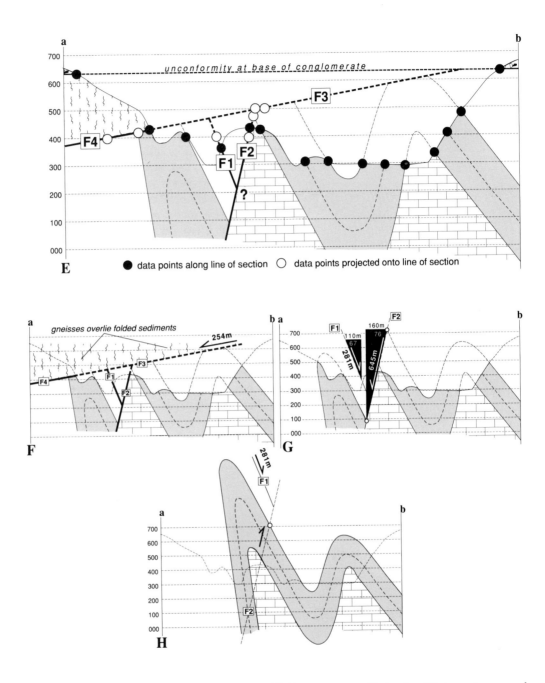

the larger displacement on **F2**, it seems most likely that **F1** is conjugate and antithetic to **F2**. They formed at more or less the same time.

From our analysis of the map we can deduce the following history of geological events from oldest to youngest:

1. Formation of gneisses.
2. Deposition of internally conformable, lower sedimentary sequence from limestone to mudstone.
3. Folding due to east–west compression.
4. East–west crustal extension generates **F1** and **F2**.
5. First movement on thrust fault (**F3–F4**).
6. Emplacement of dykes.
7. Second movement on thrust fault due to WSW–ENE compression.
8. Uplift and erosion.
9. Deposition of conglomerate.

Exercise 14.0.4

In this exercise we have no direct evidence for topographic height except that given by the topographic profile. Indirect evidence is however given by the courses of the rivers and streams. We are also given the stratigraphic succession. Thus appraisal of the attitudes of contacts and faults has to rely on interpretation of outcrop shape relative to topography as well as the direct measurements of dip and strike of bedding.

The volcanic rocks and conglomerate are the youngest formations, apparently occupying only the higher ground. Their highly sinuous outcrop pattern in relation to the streams is very different from that of the slates, quartzite and sandstone and suggests very low dips. In many places the outcrops of these younger rocks cut across faults and contacts in the older rocks, demonstrating the presence of an unconformity (diagram **A**). Note that the volcanic rocks, whilst having a similar attitude, extend further to the east than the conglomerate, a situation known as **overlap** or **onlap**.

The attitudes of the faults can be estimated from the relationships between outcrop shape and topography (i.e. 'veeing' in the valleys; see Chapter 4). Thus faults 2, 3 and 4 (diagram **A**) are vertical; 5, 6, 7, 8, 9 and 13 dip to the SW at moderate angles and have a general NW–SE strike; 1, 11 and 12 dip steeply to the SW and strike again NW–SE; the dip and strike of fault 10 cannot be determined directly. Note also that fault 1 cuts 2, 3, 7 and 8; 2 cuts 5 and 6; 3 cuts 8, 9, 11 and 12; and fault 4 cuts 9, 10 and 11 (diagram **A**).

Assessment of the measurements of dip and strike of bedding in the older sequence of sediments shows that they are folded into an overturned anticline and syncline (diagram **B**). Cleavage/bedding relationships show that the cleavage relates to the folds—it is axial planar—and that the shared limb of the folds is overturned towards the NE. The asymmetry of the minor folds, when viewed in the same direction, also relates to the presence of major folds and shows these to have low plunges to the NW and/or SE (refer back to Chapter 12). In the older rocks lying to the SW of fault 1, cleavage consistently dips more

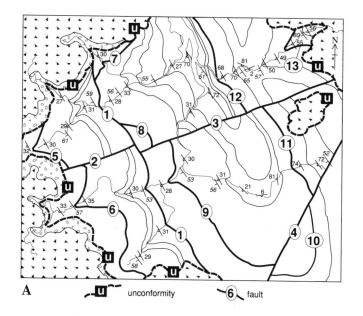

A   **u** unconformity   6 fault

steeply than bedding and minor folds are Z-shaped when viewed from the SE. Therefore structurally this area lies on the uninverted limb of a major fold—possibly the main anticline to the NE.

In the north-eastern part of the area, dips of bedding, cleavage/bedding relationships, minor fold asymmetry and outcrop shape reveal the presence of a syncline (diagram **B**). Dips of bedding in the hinge zones of the folds also suggest a very low plunge to the SE.

Consideration of the attitudes of the faults and their cross-cutting relationships suggests that we can correlate some of them (diagram **C**). Fault 1 dips more steeply than the bedding and is the youngest. The vertical fault 2 (diagram **C**) is displaced by 1 but itself cuts 4 and 5, which appear to have dips close to that of the bedding in the older sediments; the outcrop of fault 5 follows that of the bedding around the anticline (diagram **C**) and it is therefore folded. Fault 6 again has a similar dip to the bedding and could therefore be the same as 5 folded around the syncline; like 5 it lies beneath the sandstone (diagram **C**).

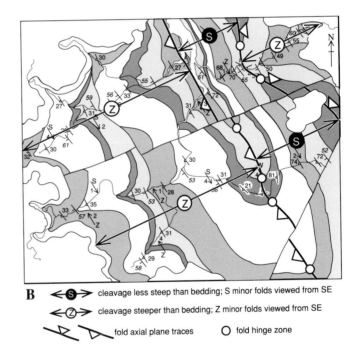

B cleavage less steep than bedding; S minor folds viewed from SE

cleavage steeper than bedding; Z minor folds viewed from SE

fold axial plane traces   ○ fold hinge zone

Because we are given the stratigraphic succession we can determine the throw on the faults (the downthrow side will be towards the side containing the younger rocks, diagram **C**; refer back to Chapter 8). The downthrow on faults 1 and 4 is consistently to the NE and as both faults dip to the SW this may suggest up-dip slip, i.e. that they are thrust or reverse faults. Fault 5 has a downthrow towards the core of the anticline with the oldest rocks sitting above the fault suggesting that it is a folded reverse or thrust fault. Fault 6 dips to the SW and carries older rocks above younger—again implying a thrust or reverse fault displacement. In contrast faults 2 and 3 do not show consistency of throw and both cause a horizontal shift of the folds (diagrams **B** and **C**). Because the senses of shift on these faults are consistent along their lengths, and because there is no obvious change in the level of erosion of the folds across them, they appear to be strike-slip faults (refer back to Chapter 13). In this analysis of the faults it is however important to realise that except for faults 2 and 3 we have no positive evidence for directions of slip; there are no displaced linear features or vertical planar structures of the right orientation to permit accurate assessment.

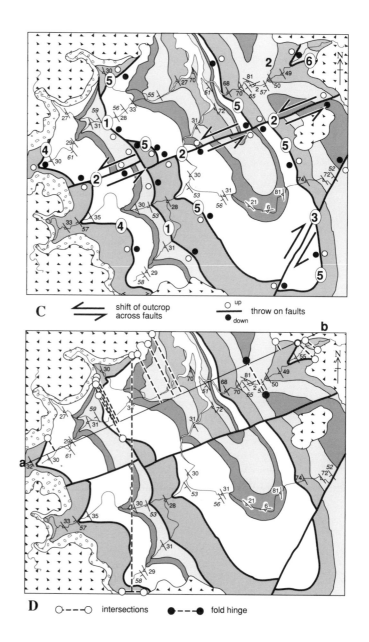

C shift of outcrop across faults   ○ up / ● down throw on faults

D ○--○ intersections   ●--● fold hinge

With this degree of understanding of the possible structure of the area we can now attempt to draw a cross-section. In doing this we can use not only information along the line of section but also data projected onto the line of section (diagram **D**). Note in diagram **D** that we are justified in projecting those lines of intersection between the various planar structures for which we have only one data point, because we can judge the general attitudes of the unconformity and faults and we are given the dip and strike of the sediments. Diagram **E** shows the data upon which we can base the section, and diagram **F**, the completed section.

The folds in the cross-section (diagram **F**) are drawn with straight limbs because on the map we can see that the dips of the bedding on the fold limbs are consistent. The outcrop of the hinge zone of the anticline is rounded whereas that of the syncline is more angular, and this is reflected on the section. However, below the level of fault 5 in the syncline we can only estimate the fold geometry as this level is not exposed in the map area.

Diagram **G** illustrates what the section would have been before deposition of the younger rocks and before movement on fault 1. The change in thickness of the grey slate from **a** to **b** to **c** suggests that faults 4 and 5 may be the same and that the thrust changes level in the stratigraphic succession across the area; it climbs up section from NE to SW. The effects of the folding are removed in diagram **H** and we can see how these changes of thickness could relate to thrust ramps with overthrusting towards the SW.

We can now summarise the tectonic history of the area:

1. Deposition of the older sediments from sandstone to grey slate (originally a mudstone or siltstone).

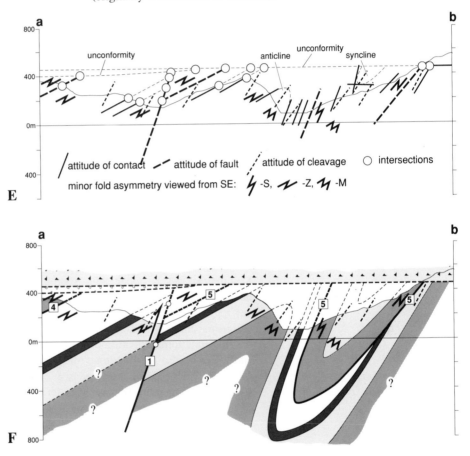

2. Overthrusting towards the SW forms fault 5; exact displacement unknown but a minimum of 5600 m if we assume that the slip direction was in the plane of the section (calculated by measuring the length of the fault in diagram **H**—before movement, the sandstones in the hanging-wall would have been continuous with those in the foot-wall).
3. NE–SW compression under elevated temperature and pressure causes folding and formation of cleavage.
4. Continued or renewed NE–SW compression causes formation of strike-slip faults 2 and 3; slip on fault 2 is sinistral (left-lateral) by about 200 m (the amount of offset of the anticline); on 3 it is dextral (right-lateral), again by 200 m. The two faults comprise a conjugate set formed at the same time.
5. Further NE–SW compression leads to movement along fault 1 with possibly reverse slip of about 380 m (measured on cross-section).
6. Uplift and erosion.

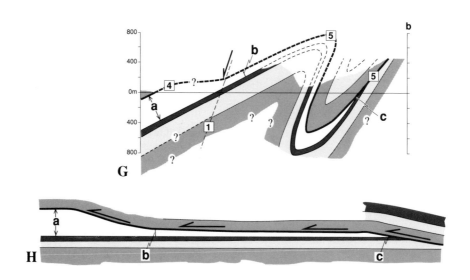

7. Deposition of conglomerate unconformably over older rocks and structures; original extent eastwards not known.

8. Formation of volcanic rocks, possibly conformable with conglomerates but overlapping them to the east.

Exercise 14.0.5

Despite the lack of topographic contours and spot heights, we can quickly deduce from the outcrop patterns in this map (in particular the repetition of outcrops on the mountainsides across the inlets) that a flat-lying sequence of sediments and faults is cut by a dyke and a steeply dipping fault. The conglomerate forms the hilltop in the north unconformably overlying the dyke and therefore also the limestone (diagram **A**).

Dips of bedding confirm not only the low-lying attitudes of the sedimentary formations but also of faults 2, 3 and 5 (their outcrops follow those of the lithological contacts and therefore they must have similar dips). Dips of bedding change systematically across the area from west to east such that a broad antiform crosses the area from north to south. As well as this open fold, tight to isoclinal folds are picked out by the turn-around of the outcrops of the limestone on the valley sides in the west and'north-west of the area (diagram **A**). By locating the hinges of these folds and considering the dip and strike of the rocks, we can see that these are outcrops of the same north–south trending fold whose hinge must be sub-horizontal. The same fold can be located in the south-western corner of the map by projecting the hinge lines further to the south (diagram **A**). Note that the lower limb of this fold extends across the area to the east but that the upper limb is truncated upwards by faults 2 and 3 (diagram **A**).

'Veeing' into the valley in the east demonstrates that the dyke dips at a moderate angle to the N–E; it is cut by fault 1 but cuts fault 5. Fault 1 dips steeply to the NW striking NE–SW (diagram **A**) and is the youngest fault in the area. Fault 4 strikes NE–SW and dips to the SE at a moderate to steep angle. There is no evidence for its relationship to the dyke.

The stratigraphic succession tells us that there is only one limestone, siltstone, etc., and therefore that on the map the repetition of the outcrops of the sedimentary formations must be due to folding and/or faulting. We can investigate the effects of folding further by noting the relationships between bedding and cleavage and the way in which the sediments are stacked on top of

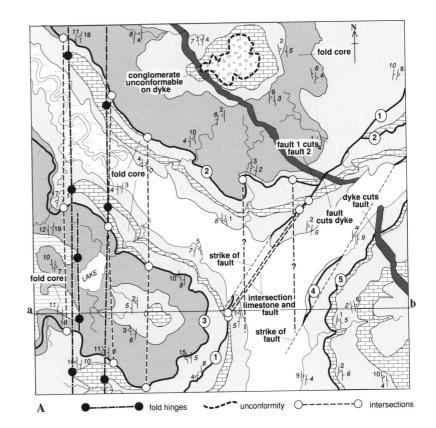

A ●━·━·━● fold hinges ━ ━ ━ unconformity ○─ ─ ─ ─○ intersections

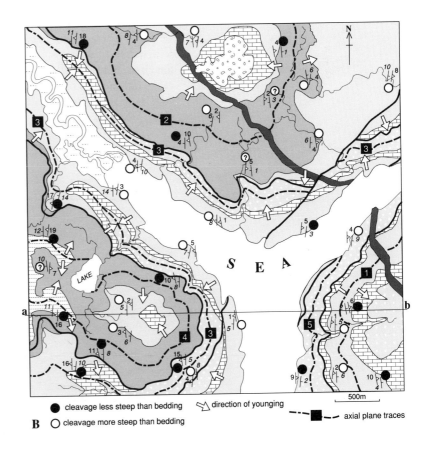

B ● cleavage less steep than bedding ⇦ direction of younging ━·■·━· axial plane traces
○ cleavage more steep than bedding

each other, i.e. whether they get younger upwards or downwards (diagram **B**). In the south-eastern corner the limestone occurs both above and below the siltstone; the sequence above is in stratigraphic order and is getting younger upwards, whilst that below is inverted (arrows in diagram **B**). As no fault is indicated, the outcrop of the siltstone must contain the axial plane of an isoclinal fold. Likewise in the north, the sandstone both under- and over–lies the siltstone, indicating the presence of another isoclinal fold. In both cases these changes in younging directions coincide with changes in the relationships between cleavage and bedding, thus confirming the presence of fold closures. Note, however, that neither are inverted rocks everywhere characterised by the dip of cleavage being steeper than bedding, nor uninverted rocks by the opposite relationship. This arises because, although the cleavage is axial planar to the large isoclinal folds, the attitudes of cleavage and bedding are further affected by the large N–S open fold (see later). Despite this, by using 'younging' and cleavage/bedding relationships we can identify the axial plane traces of five major, tight to isoclinal, flat-lying folds (diagram **B**).

From our understanding of the structure so far, by using data gained directly along the line of section and by projecting other data onto this line (diagram **A**), we can begin to construct a cross-section (diagram **C**).

Section **C** shows the data plotted onto the topographic profile. Whilst we do not know directly the height of some of the data points, e.g. the projected points and the dips, we do know where *along* the section the intersections and dips should lie. The intersections at ground surface do however provide constraints as to their approximate heights because, for example, a dip on the map in a particular formation must also lie in that formation on the section. The intersections of the limestone and sandstone contacts with fault 2, lying between the faults 1 and 4 in section **C**, are projected onto the section from the ground to the north-west of fault 1.

The dip of faults 1 and 2 can be determined reasonably accurately because we know where on the section they lie at sea level and where they cut the section line (diagram **A**). The shift on fault 1 is given by the offset of the limestone and therefore, knowing this, we can determine the positions of other geological contacts between faults 1 and 4 (section **D**) and, in turn, the shift on fault 4; it appears to downthrow to the east.

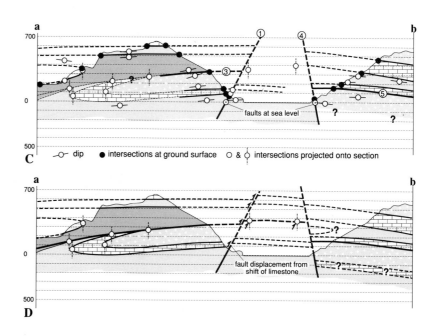

Section **E** shows the positions of the axial surfaces of the major, flat-lying folds and suggests that we can correlate the folds and the flat-lying faults across the area. Thus folds 3 and 5 are the same (fold 2 in the section) as are 1, 2 and 4 (fold 1 in the section). Likewise faults 2, 3 and 5 in diagram **B** are the same structure (fault **a** in diagram **E**). This is more clearly seen in diagram **F**, where the effects of faults 1 and 4 have been removed. Evidently a pair of flat-lying folds are displaced from **x** to **x** along the flat-lying fault. The fault emplaces older rocks on younger and is thus most likely to be a thrust fault. We have no direct evidence for its direction of slip but were it related to the compression that caused the folding it would have been E–W.

Fault 1 downthrows to the NW but this could be caused by dip- or oblique-slip; fault 4 downthrows to the SE, but again this could be due to oblique- or dip-slip. However, both of these faults involve extension and probably formed as a conjugate system during NW–SE crustal extension.

In summary the history of the area is:

1. Apparently conformable deposition of the lavas to sandstones.
2. Folding due to E–W compression; formation of cleavage and overturned, isoclinal, flat-lying folds.
3. Thrusting due to E–W compression; possible displacement of about 2150 m.
4. Formation of open N–S antiform by E–W compression.
5. Intrusion of dolerite dyke.
6.* Uplift and erosion.
7.* Deposition of conglomerate unconformably over older rocks and structures.
8.* NW–SE crustal extension forms faults 1 and 4.

It is not clear from the map whether the conglomerates were deposited before or after faulting on 1 and 4.

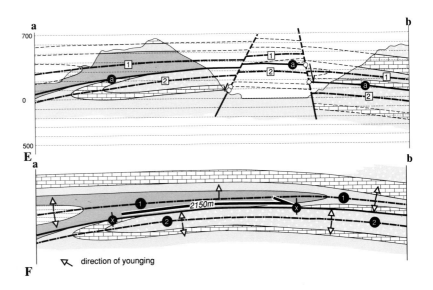

direction of younging

Exercise 14.0.6

In this map again we have limited topographic information and are not given the stratigraphic succession. Despite this we can quickly establish that the sandstone and conglomerate form the high ground and overlie a complex of cleaved sediments. Cross-cutting relationships show the base of the sandstone to be unconformable. The sandstone dips at 4–5° to the NW (given by dip and strike data). Calculation of the dip of the unconformity using spot heights and solving the three-point problems (refer back to Chapter 6) shows this to dip at about 7° to the NW (diagram **A**). This difference in attitude may be due to local variation in dip or to deposition of the sandstone on a sloping palaeo-topographic surface against which the sandstone is banked up.

The base of the overlying limestone to the east of fault 3, both in the north and the south of the area, lies at about 2500 m and is thus horizontal. To the west it is again horizontal but lies at 3000 m, a change of height presumably attributable to movement on fault 3. Again, cross-cutting relationships show the limestone to be unconformable (diagram **A**).

Using appropriate spot heights and by solving the three-point problem, we can calculate the attitude of fault 3; it dips at about 33° to just north of east (diagram **B**). It downthrows to the east. By projecting the line of intersection of the two unconformities onto the fault plane (diagram **B**), we can determine the amount and direction of slip (refer back to Chapter 8). The fault is extensional, with a slip direction only slightly oblique to its dip.

The rocks lying below the unconformities are all cleaved. The outcrop patterns in relation to topography, together with the dips of bedding, reveal the presence of three, angular, major folds to which the cleavage is axial planar (diagram **C**). Because the strike directions of bedding and cleavage are consistently NNE–SSW, the folds must be essentially cylindrical with horizontal hinges.

In the east the antiform is cored by black slate succeeded upwards, in turn, by siltstone, grey slate and finally conglomerate. West of fault 3,

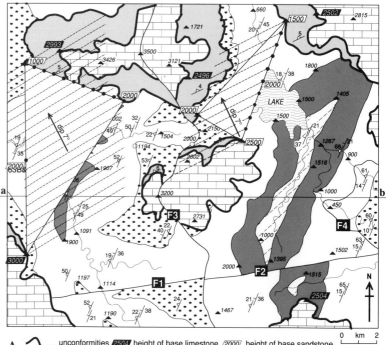

A ⌇⌇ unconformities ▨2504▨ height of base limestone ▨2000▨ height of base sandstone

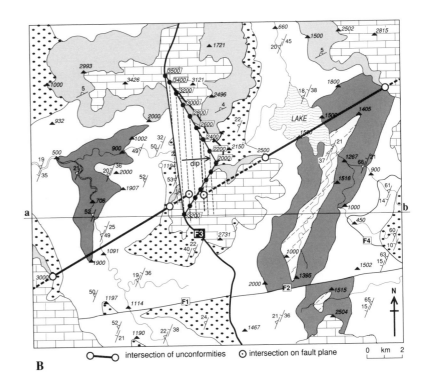

intersection of unconformities ⊙ intersection on fault plane

0 km 2

B

siltstones, in the core of the antiform, are overlain by grey slates succeeded by conglomerates. Note that there is no evidence for an unconformity at the base of the conglomerate; there are no cross-cutting relationships and the conglomerate dips are concordant with those in the other cleaved sediments. Thus the vertical (stratigraphic?) succession would appear to be:

Youngest: Limestone

------------------------------------- *Unconformity*

Sandstone

------------------------------------- *Unconformity*

Conglomerate

Grey slate (originally mudstone)

Siltstone

Oldest: Black slate (originally mudstone)

As we have seen, fault 3 dips at 33° to the east and is essentially a normal fault with a vertical throw of 500 m (the difference in height of the base of the horizontal limestone on either side). Faults 1 and 2 are vertical and cause the fold hinges to shift horizontally by about 2600 m (diagram C). As there is no discernible change in height of the hinges across these faults, they must be left-lateral (sinistral) strike-slip faults and were, before movement on fault 3, contiguous, i.e. the same fault.

Judging by its outcrop shape relative to topography, fault 4 dips at a shallow angle towards the NW. It cuts the steeply dipping rocks of the overturned limb of the anticline. Because the rocks are overturned it is not possible to determine the downthrow side of the fault, nor its nature, from the map.

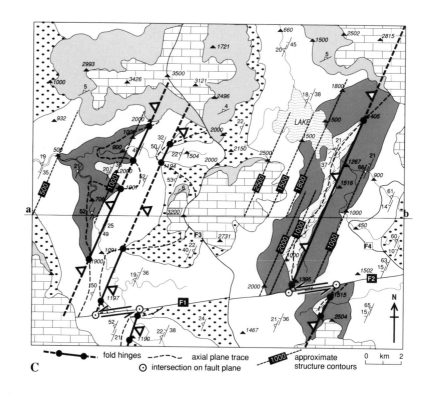

fold hinges ⚫━━⚫ ----- axial plane trace ▬▬1000 approximate structure contours

⊙ intersection on fault plane

0   km   2

C

In the older rocks, the consistency of strike allows us to construct structure contours where the outcrops of contacts coincide with spot heights (diagram **C**). These can be used together with other projected data, and those gained from the line of section, in constructing the cross-section (diagram **D**). In section **E** the folds are drawn in through the relevant data points; they are angular in shape because limb dips are regular (see map). The relationships of the faults 3 and 4 cannot be determined from the map, so their positions at depth on the section are only tentatively drawn in.

The completed section (diagram **F**) reveals about 1200 m down-dip displacement of the base of the limestone across fault 3 but only 800 m for that of the conglomerate. The fault was therefore active before or during limestone deposition as well as after. By drawing the base of the conglomerate around the eastern antiform (after taking out the effects of post-limestone movement on

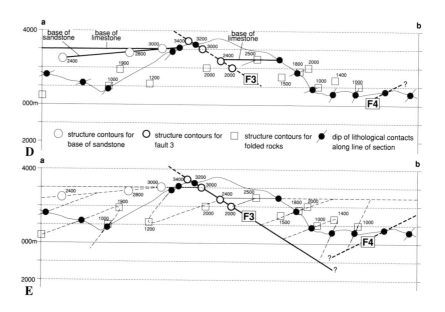

structure contours for base of sandstone   structure contours for fault 3   structure contours for folded rocks   dip of lithological contacts along line of section

D

E

fault 3, diagram **G**) we can deduce that fault 4 causes down-dip offset and may therefore be an extensional fault. In doing this we are however assuming no change in the shape of the fold along its axial surface.

In summary, the structural history of the area is as follows:

1. Conformable deposition of mudstone (now black slate) to conglomerate.
2. Folding and cleavage formation at depth in the crust, due to WNW–ESE compression; overturning to the ESE.
3. Formation of strike-slip fault (1–2) due to NW–SE compression.
4. Movement on fault 3.
5. Uplift and erosion.
6. Deposition of sandstone banking up against an inclined erosion surface.
7. Earth movements causing tilting of sandstone.
8. Uplift and erosion.
9. Deposition of limestone unconformably over whole area; fault movement?
10. Extensional movement on fault 3 (and fault 4?) due to almost E–W crustal extension.

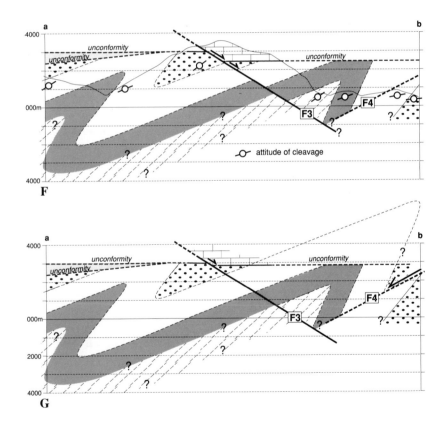

Exercise 14.0.7

A quick appraisal of the map and key reveals that a younger sequence of flat-lying conglomerate and lava unconformably overlies cleaved sediments and volcanics, with steep to moderate dips to the SW. Three vertical faults occur, one of which cuts the younger sequence. Closed outcrops of grey sandstone in the south-west of the area together with changes in the dip of bedding suggest the presence of folds. Low-angled faults dipping to the SW are present between the vertical faults. The vertical and/or stratigraphic sequence in the cleaved sediments is not at once obvious.

Thus we can rapidly summarise the main features of the map, which is a complex one, and this can guide our more detailed analysis.

For ease of description the faults have been labelled in diagram **A** and the area divided into four sub-areas, **A–B–C–D**, bounded by the vertical faults. Throughout the map area, truncation of boundaries in the cleaved sediments and differences in dip show that the conglomerate unconformably overlies the older rocks, and fault **g**. Likewise the overlying lavas are unconformable in the SW but could be conformable with the conglomerates in the NE. The lavas may overlap the conglomerate towards the SW. The lavas and therefore all older rocks are intruded by a vertical basalt dyke trending WNW to ESE.

We can determine the vertical (stratigraphic?) sequence for the older sequence by imagining ourselves walking across the area taking note of the positions of the various formations relative to their dip. Thus if we start in the grey sandstones in sub-area **A**, to the SW of fault **c** (locations **i** in diagram **A**), we can walk to north or south through the siltstones, then red sandstones, and into the volcanic ashes (locations **iv**). Along the southerly traverse the rocks dip consistently to the SW at shallow to moderate angles, whilst to the north they dip steeply to the SW. To the north the cleavage dips less steeply than bedding, but to the south more steeply. Thus localities **i** are in the core of an overturned antiform and, as would be expected, the sequence of rock formations is repeated on both limbs of the fold. The lowest (and possibly oldest) formation is therefore the grey sandstone and the original vertical sequence must have been: grey sandstone–siltstone–red sandstone–volcanic ashes. Similar considerations show this to be the vertical sequence for the ground in each of the sub-areas lying to the SW of the low-angled faults and nowhere is there any evidence to show that this sequence is not internally conformable.

A similar traverse in sub-area **A** to the NE of the low-angled fault **c** shows repetition of succession across a synform whose core is occupied by red sandstones (locality **iii** in diagram **A**). Note here, however, that on the northern limb of the fold the siltstones are underlain by mudstones, not grey sandstones (locality **i**). As there are no indications of unconformable relationships or faults to explain this situation, the evidence suggests that there is a lateral facies

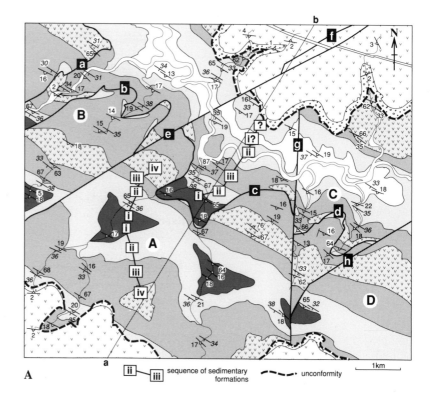

A

ii — iii sequence of sedimentary formations   – – – · unconformity

1km

change from mudstone to grey sandstone.

Shifts of outcrops across the vertical faults and repetition of outcrop pattern across the valley in the NW corner of the map suggest that faults **a**, **b**, **c** and **d** may be segments of the same fault which, because of its low south-westerly dip, underlies all the area to the south-west of its outcrop. Thus, in considering the fold structures affecting the older sedimentary sequence, we should look firstly at the ground to the south-west of this fault (**a** in diagram **B**) and then to the north-east (or vice versa).

From repetitions of sequence, dip of bedding (deduced from readings on the map and relationships between outcrop shape and topographic indicators) and cleavage/bedding relationships, a number of major overturned folds can be located, as shown in diagram **B**. These are numbered according to possible correlations across the vertical faults **e**, **g**, and **h**. The folds are essentially cylindrical with sub-horizontal hinges.

A similar analysis of the older sediments lying to the north of fault **a** shows the presence of further folds which must underlie the fault (diagram **C**).

Truncation of the structures underlying fault **a** is indicated in diagram **D**, as is the intersection between the lavas and conglomerate. These will be used later in constructing the cross-section.

As we have seen in previous exercises, the outcrop patterns of folds and location of their hinges can be very useful in determining slip on faults. In this exercise accurate location of fold hinges both above and below fault **a** allows us to determine the amounts of slip on faults **e** and **g**. Location of their axial surfaces on the cross-section further permits assessment of the possible slip on fault **a**.

As the fold hinges can be shown to be sub-horizontal, we are justified in projecting these onto the outcrops of faults **e** and **g** as in diagram **E**. There are no indications from the outcrop patterns of the folds that vertical movements across the faults are important; the folds are consistently cored by the same rock formations across the map. Consequently the obvious shift of the fold hinges must be attributable to strike-slip; fault **e** is a left-lateral (sinistral) strike-slip

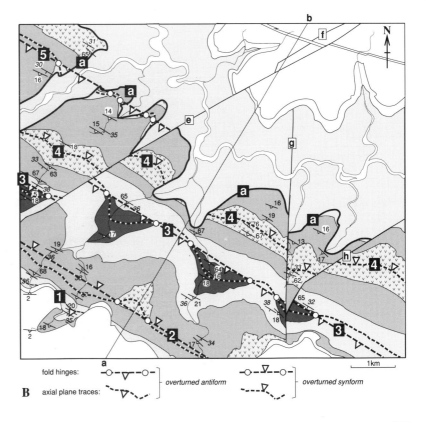

fold hinges: ○—▽—○ overturned antiform ○—▽—○ overturned synform

B axial plane traces: ▽ ── ▽

1km

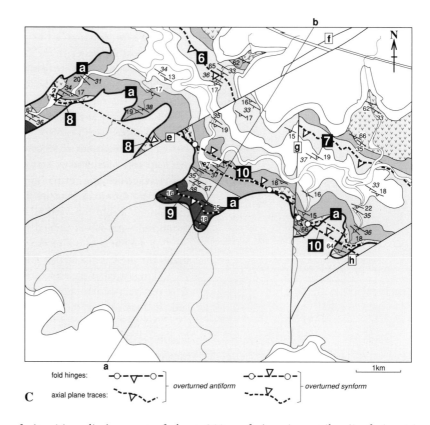

fold hinges: —○─▽─○─ ⟩ *overturned antiform*   —○─▽─○─ ⟩ *overturned synform*

C axial plane traces: ---▽-··-··- ⟩   ---▽-··- ⟩

fault with a displacement of about 850 m; fault **g** is a strike-slip fault with about 200 m of right-lateral (dextral) slip. Note, however, that fault **e** causes only 150 m of slip where it displaces the basalt dyke. The fault moved both before dyke emplacement (by 700 m) and after (by 150 m). Slip on fault **g** occurred before deposition of the conglomerate as this unconformably overlies it.

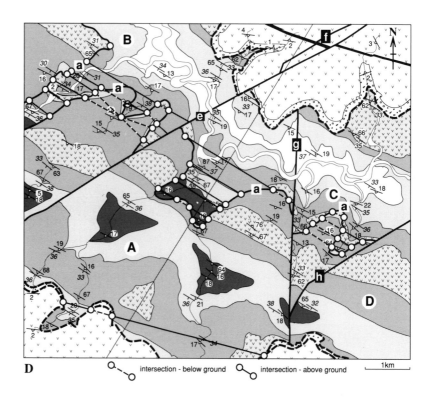

D   ○--○ intersection - below ground   ○—○ intersection - above ground   1km

Fault **h** causes a shift in the position of the hinge of fold 3 and the axial surface of fold 4 (diagram **E**). There is no evidence for substantial vertical movements across the fault and it is therefore a left-lateral strike-slip fault with a displacement of about 500 m. Faults **e**, **g** and **h** form a conjugate set whose orientation suggests maximum compression acting NE–SW and causing fracturing of the rocks both before and after deposition of the younger sedimentary/volcanic sequence.

As in previous exercises, when drawing the cross-section, we make use of data from as much of the map area as is justified by its structural complexity, not just those along the line of section. Thus we can build up the sections as seen in diagrams **F** and **G** and arrive at the final section (diagram **H**).

It is clear in diagram **H** that the folds above and below the low-angled fault have the same geometry; they are angular and asymmetric with similar axial plane and limb dips. From the map we know that their hinges all trend NW–SE and are sub-horizontal. They are likely therefore to have formed at the same time and have been displaced by the low-angled fault. Because the folds have the same hinge attitudes above and below the fault there has been no rotation across the fault plane. Movement therefore involved either strike-, dip- or oblique-slip. If you have constructed your section as accurately as possible, it should be apparent that the shapes of folds above the fault can be matched with those below, as shown in diagram **I**. This geometric fit, whilst not providing absolute proof, indicates strongly that fault movement was accomplished by slip in the line of section, i.e. dip-slip. The fault is an extensional normal fault with about 1250 m of dip-slip. It clearly pre-dates formation of the vertical strike-slip faults and therefore is earlier than deposition of the conglomerate.

The history of the area can now be summarised:

Oldest: 1. Deposition of older sequence.
        2. Folding and cleavage formation by compression from the SW.
        3. Extensional fault **a** formed.

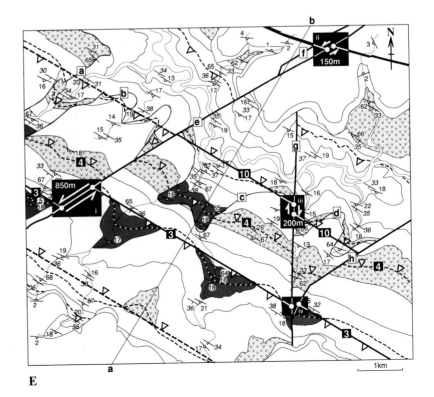

E

4. Formation of faults **e**, **g** and **h**—NE–SW compression.
5. Uplift and erosion.
6. Deposition of conglomerate followed by extrusion of lava. Lava overlaps conglomerate.
7. Renewed movement on fault **e** (**f**).
8. Intrusion of basalt dyke.

Youngest: 9. Renewed movement on fault **e**.

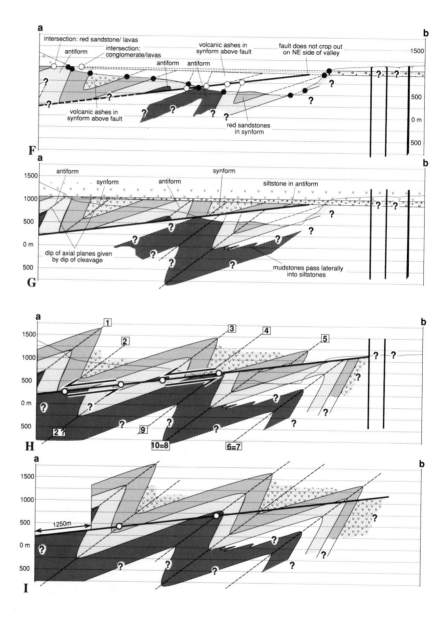

Exercise 14.0.8

Consideration of cross-cutting relationships, dips of bedding and outcrop shapes reveals unconformities at the base of the limestones, the conglomerates, siltstones and sandstones (diagram **A**). The limestones occupy the highest ground and unconformably overlie the siltstones and slates. In the north-east corner of the map the siltstones obviously unconformably overlie the sandstones. Elsewhere their relationships to the sandstones are not so obvious. In many places they are however clearly unconformable upon the slates, as are the sandstones. The conglomerates unconformably overlie all the other formations and many of the faults. On the basis of these observations a stratigraphic sequence can be suggested:

Youngest:	Conglomerates
	Limestones
	Siltstones
	Sandstones
Oldest:	Slates

The basalt occurs as vertical to steeply dipping dykes which intrude all the rocks except the siltstones, limestones and conglomerates.

All the faults have an overall north–south trend, indicating that they strike in this direction. 'Veeing' of their outcrops in relation to topography suggests the following attitudes: faults 1 and 2—moderate to shallow dips to the west; fault 3—steep dip to the west; faults 4 and 5—moderate to steep dip to the east.

In diagram **B**, intersections at unconformities and on fault planes, as well as spot heights lying at contacts and faults, are projected onto the line of section in order to allow construction of a cross-section. Some of these data points are however projected over long distances, so we have to be cautious in relying too heavily on them (slight changes in attitude may be present and could move projected points along the line of section). Bearing this in mind we can draw an approximate cross-section which allows us to assess the attitudes of the contacts and faults more accurately (diagram **B**). Note that the sandstones, siltstones and conglomerates are folded down towards the east and faults 1 and 2 are convex downwards and therefore listric.

In order to assess the nature and effects of the faults we need to determine their downthrow and displacement of markers. Faults 1 and 2 downthrow to

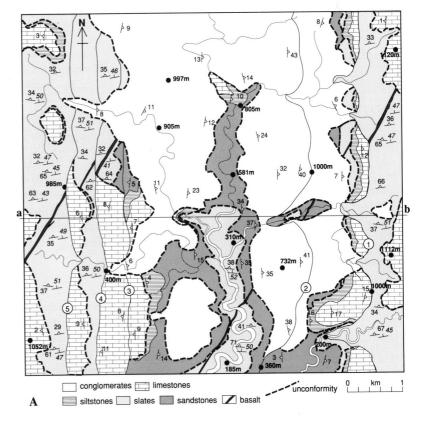

conglomerates   limestones
siltstones   slates   sandstones   basalt
unconformity
0   km   1

A

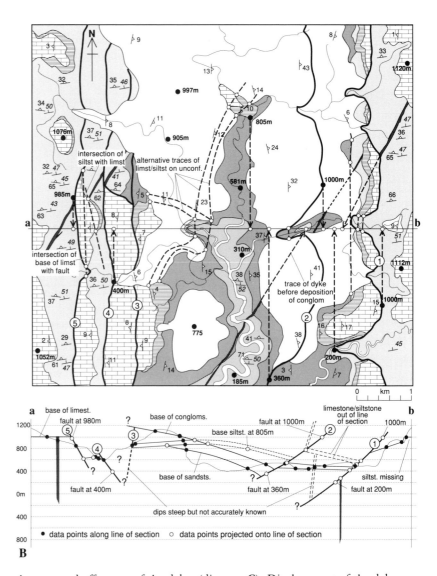

the west and offset one of the dykes (diagram **C**). Displacement of the dyke can be described by the horizontal shift in the direction of fault dip (**d**), the shift along strike (**s**) and the vertical separation (**v**) (diagram **C**). These displacements could be caused by dip-, oblique- or strike-slip. The listric shapes of the faults and, as will be discussed later, the localised deposition of the conglomerates, siltstones and sandstones do however suggest that the faults are extensional. Notice in the section of diagram **C** that offsetting of the dyke and the base of the conglomerates across fault **1** is not the same. Likewise the shifts of the dyke, the base of the sandstones and the base of the limestones across fault **2** are different—relationships which imply repeated movement on the fault planes. Because we can neither draw structure contours nor find linear features that can be projected onto the fault planes, we cannot determine precise directions and amounts of slip.

We can say little about fault **3** other than that it could be an extensional fault downthrowing to the west. Faults **4** and **5**, however, displace the intersection of the unconformities at the base of the limestones and siltstones as well as a dyke (diagram **C**). The point of intersection of this limestone/siltstone intersection with the dyke provides a feature that, before faulting, was in one place. Locating this point in each of the fault blocks thus enables us to determine the mean slip directions (diagram **C**). Fault **5** is therefore an extensional fault with slip slightly oblique to dip whereas, though extensional, slip on fault **4** was more markedly oblique.

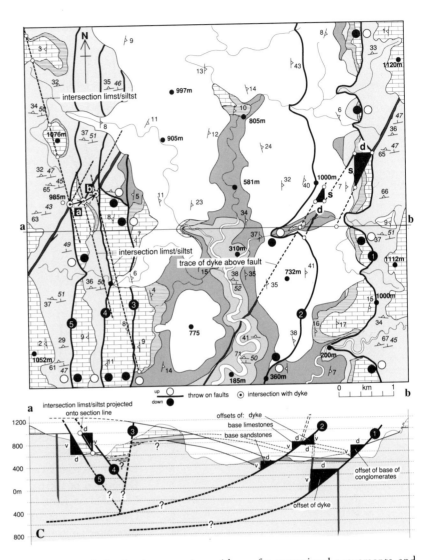

All the faults in the area give evidence for extensional movements and thus may be related. The listric geometry of faults **1** and **2** suggests that they may underlie the remaining faults (section in diagram **C**) and that the latter are due to collapse of the hanging-wall anticline as movement on faults **1** and **2** proceeded.

Diagram **D** shows the approximate locations of the axes of the hanging-wall anticline and the hinge zones of folds in the slates of the basement. Positions of the latter are indicated not only by changes in dip but also by changes in the relative dips of cleavage and bedding (refer back to Chapter 12). The cleavage must be axial planar to the folds and the folds are overturned to the SSE. The three folds in the basement to the west of fault **5** may be the same as those to the east of fault **1**. If so, their offset accords with the faults in the area being essentially normal extensional faults responsible for the development of a major rift valley running N–S. Because the base of the conglomerates is irregular, as shown by the way it infills valleys in the west of its outcrop and its irregular truncation of older formations (diagram **B**), it must form an infill of the rift at an earlier stage in its development.

We can now summarise the structural history of the area:

1. Deposition of mudstones (now slates).
2. Crustal compression, almost N–S, gives rise to major folds and cleavage; overturning to the south.
3. Uplift and erosion.

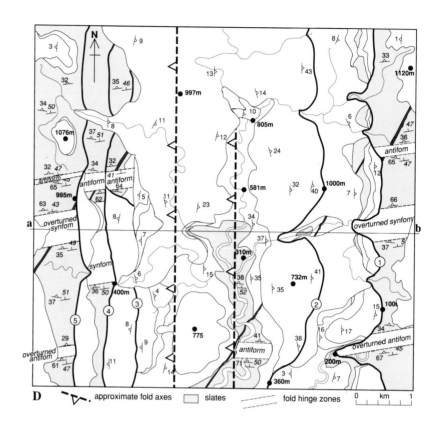

D · ⌐ꞁ approximate fold axes ▢ slates ------- fold hinge zones 0 km 1

4. Deposition of sandstones (only in central basin—proto-rift?).
5. Intrusion of dykes.
6. Earth movements cause warping and erosion.
7. Deposition of siltstones (only to the west of fault 1?).
8. Extension on fault 1.
9. Deposition of limestones across whole area.
10. Extension on faults 1 and 2 with collapse of hanging-wall anticline forming faults 3, 4 and 5.
11. Uplift and erosion.
12. Deposition of conglomerates in rift valley.
13. Further extension reactivates faults 1 and 2.

The sequence of faulting outlined above can be illustrated by constructing a series of sections which remove, in sequence, displacements on the faults (diagrams **E** to **I**). Diagram **E** shows the cross-section as seen now; **F** takes out displacement of the base of the conglomerate along faults 1 and 2 (horizontal extension = **a**′; note that the base of the sandstone is still offset); **G** restores the displacements of the base of the limestone on faults 3, 4 and 5 and removes the overlying conglomerates (total horizontal extension = **a**″); in **H** displacements of the sandstones and siltstones along faults 1 and 2 are taken out (total horizontal extension = **a**‴) and in **I** rotation, due to formation of the roll-over anticline, is removed. Note that cross-section **I** suggests that deposition of both the sandstones and siltstones could have been controlled by subsidence of a basin developed as a result of movement on fault 1 (neither of these sediments occurs to the east of fault 1 and they are apparently cut out by the limestone to the west). Alternatively the sediments originally could have spread across the whole area, were tilted by movement on fault 1, and then eroded before deposition of the limestones. The total horizontal extension of the section is given by **h** in diagram **I**.

Such restorations as these are extremely important in attempting to gain a full understanding of the structural evolution of an area.

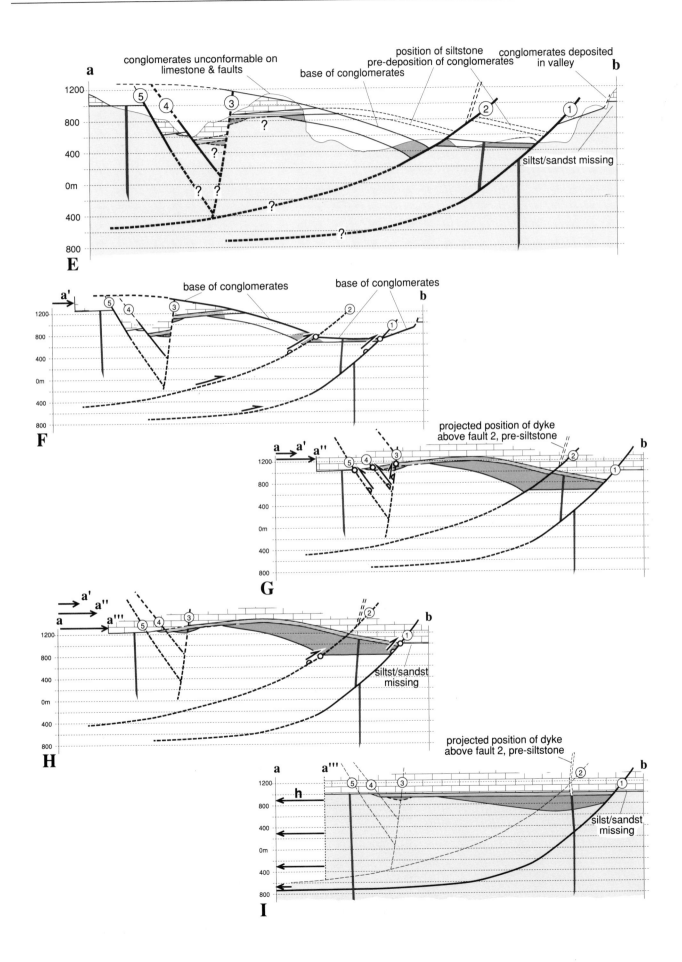

E — conglomerates unconformable on limestone & faults; position of siltstone pre-deposition of conglomerates; base of conglomerates; conglomerates deposited in valley; siltst/sandst missing

F — base of conglomerates; base of conglomerates

G — projected position of dyke above fault 2, pre-siltstone

H — siltst/sandst missing

I — projected position of dyke above fault 2, pre-siltstone; silst/sandst missing

INDEX